The Dog Listener

Learn How to Communicate
with Your Dog for
Willing Cooperation

JAN FENNELL

30043003467879

Quill
A HarperResource Book
An Imprint of HarperCollins*Publishers*

The author and publisher are grateful to the following for use of photographic material: **Insert 1: Photo #7** © Peter Orr; **Photo #9** © *Daily Mail;* **Photo #12** © Bill Forbes; **Photo #'s 14, 16** © Tracey Anne Brooks, Mission Wolf. **Insert 2: Photo #'s 18, 19** © *Scunthorpe Evening Telegraph;* **Photo #31** © *Daily Mail.*

Grateful acknowledgment is made to The National Trust for Places of Historic Interest or Natural Beauty for permission to quote from the poem "If" by Rudyard Kipling.

THE DOG LISTENER. Copyright © 2000, 2004 by Jan Fennell; foreword copyright © 2000 Monty Roberts. All rights reserved. Printed in the United States of America. No part of this book may be used or reproduced in any manner whatsoever without written permission except in the case of brief quotations embodied in critical articles and reviews. For information address HarperCollins Publishers Inc., 10 East 53rd Street, New York, NY 10022.

HarperCollins books may be purchased for educational, business, or sales promotional use. For information please write: Special Markets Department, HarperCollins Publishers Inc., 10 East 53rd Street, New York, NY 10022.

The 30-Day Training Guide was first published 2002, as *The Practical Dog Listener* by HarperCollins Publishers in the United Kingdom. Photography in *The 30-Day Training Guide* by Stephen Daniels.

Library of Congress Cataloging-in-Publication Data has been applied for.

ISBN 0-06-008946-6

04 05 06 07 08 WBC/RRD 10 9 8 7 6 5 4 3 2 1

Caution

It is important to say here that my method cannot remove the aggressive tendencies of any dog. Certain breeds have been raised specifically for the purpose of fighting, and my methods will never be able to alter their potentially savage nature. What my method can do is allow people to manage their dogs so that this aggressive instinct is never called upon. Please exercise the greatest of caution when working with such dogs.

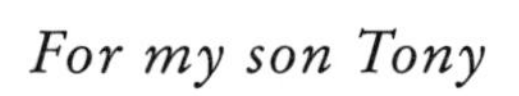

For my son Tony

Contents

Foreword

Dogs have played an important part in my life. My wife, Pat, and our family have had several over the years that were loving companions and important members of our family. It has been another wonderful creature that has dominated my career, however. I have spent my life developing—and often defending—the method I have discovered for communicating with the horse.

The appetite the dog world has for my ideas has been obvious throughout this time. Wherever I may be in the world, there are invariably four times as many dog owners and trainers as there are horse trainers at my demonstrations. Almost to a person, they have strong, positive comments to make about my method.

Given my time all over again, I would relish the challenge of adapting my ideas and taking them into the canine world. As it is, though, I have more than enough to keep me occupied, developing and sharing my own discipline. Fortunately, in the last few years, I have become aware of a talented dog trainer who, inspired by my method, has undertaken the task already.

It was with a warm heart that I first learned of the work Jan Fennell has been doing in England. I have been lucky enough to meet her there and she has related much that reminds me of my own earlier experiences. Like me, Jan feels a deep sense of injustice at the way man has sometimes maltreated an animal he claims to call his friend. She also passionately believes that violence has no place in our relationship with animals. Her dream, too, is a world in which all species live in peace.

As with me, Jan has been slow to summon the courage to tell her story. I dragged my feet for years before I wrote my first book, *The Man Who Listens to Horses*. Jan has been just as careful in waiting to put her ideas into print. She now feels confident in her experience and is ready to share her remarkable work with a wider audience.

As she does so, I wish her and her ideas well. I am sure there will be those who will assail her. If my experience has taught me anything, it is that human nature has an almost limitless capacity for negativity. Each of us should be aware that for every grain of negative within the human community, there is a mountain of positive waiting for us among animals. We should also note that for every negative, however, there are literally hundreds thirsting for a better way to deal with man's best friends.

I am proud to think that by sticking to my beliefs I have helped make the world a better place for the horse and, hopefully, for people too. I hope this book can achieve the same for another very special creature, the dog.

MONTY ROBERTS, *California, March 2000*

The Dog Listener

Introduction

I am a great believer in learning from the mistakes we make in life. I should be. I have made more than enough of my own, in my relations with humans as well as dogs. Of all the lessons the latter have taught me, none was as painful as that I received in the winter of 1972. It seems to me fitting that I should begin with the tragedy of Purdey. For reasons that will soon become apparent, her story is inseparable from my own.

At the time, I was married and was raising my two young children, my daughter, Ellie, born that February, and Tony, then two-and-a-half. We were living as a family in London but had just decided to move to the countryside, and a small village in Lincolnshire, in the heart of England. Like so many people drawn to the rural life, we were all looking forward to going on long country walks and decided we would like a canine companion to take with us. Rather than buying a new puppy, we thought we'd rescue a dog. We liked the idea of giving a home to an animal that had had a raw deal, so off we trundled to the RSPCA and saw this rather sweet, six-month-old, black and white, cross Border collie-whippet. We took her home, where we decided to call her Purdey.

She was not the first dog in my life. That had been Shane, a magnificent, tricolored Border collie I had been given by my father when I was a thirteen-year-old girl growing up in Fulham, west London. I had always loved dogs and, as a little girl, had invented an imaginary one called Lady. I remember my grandmother indulging me by talking to my fictional friend with me. I think I saw dogs then, as I do now, as objects of unquestioning love and total loyalty—qualities that are hard to find in humans. Shane's arrival in our family had only confirmed my feelings.

I trained Shane with my father, according to the technique Dad had used himself in raising his dogs as a young boy. Dad was a gentle man, but he was also determined the dog was going to do what we said. If Shane did something wrong, he got a tap on the nose or a smack on his bottom. But I got a smack on the bottom too and I thought it was OK, particularly as Shane was an extremely smart creature and seemed to understand what we wanted. I can still remember the pride I used to feel at taking him onto Putney Heath and Wimbledon Common on the Number 74 bus. Shane would sit by my side without a lead, behaving impeccably all the time. He was a super dog.

If something works, you go along with it. You don't mend what isn't broken, as they say. So when we got Purdey, I decided to apply the same method as I had with Shane, teaching her the difference between right and wrong with a mixture of love, affection, and, where necessary, force.

At first this method seemed to work for Purdey too. She behaved well and fitted easily into the family in London. The problems started when we eventually moved to Lincolnshire that September. Our new home could not have presented a greater contrast to noisy, over-populated London. We lived in a small, isolated village. There were no street lights, the buses only ran twice a week, and it was a four-mile hike to the nearest shop. I remember when I was a toddler I had been taken to the seaside for the first time. I took one look at the sea and ran back up the hill away from it. My expression as a three-year-old was "too big enough" and, if she could have spoken, I'm sure that's what Purdey would have said about her new home. It seemed like everything was too big enough.

Soon after we arrived, Purdey began to behave in a way that I thought then was odd and not a little bit worrying. She would run off into the countryside, disappear for hours, then come back obviously having had a great time somewhere. She was also hyperactive and seemed to be wound up by the slightest thing or sound. She followed me absolutely everywhere I went, which was a nuisance when I had the two small children. I wasn't happy about her roaming the countryside like this. We all have a responsibility to make sure our dogs don't cause danger or become a nuisance to others. But I decided that I had taken this dog on and I was going to stick with her. I owed it to her to help her settle, and that's what I hoped to do. Events, however, soon overtook me.

The first inkling I got that something was wrong was when a local farmer came to see me. He told me in no uncertain terms that if I did not keep this dog under control he was going to shoot her. I was devastated, of course, but I also saw his point because he had livestock and Purdey was obviously running around and worrying the animals. So we put her in the huge, 200-foot garden we had, slipped a rope on her collar, and attached it to the washing line so she could go no farther. But she still ran off whenever she could.

Matters took a turn for the worse one cold winter's morning just before Christmas. I had come downstairs with the children and was going through our usual start-of-the-day routine. Purdey was frantically charging around as she always did first thing in the morning. I remember Ellie was crawling around on the floor, while Tony was playing the "little helper," sorting out a pile of clothes I had in the sitting room. I went into the kitchen, which led directly off the sitting room, to collect their drinks, when I heard a loud crash. I will never forget what I saw when I looked around. The dog had jumped up at Tony and jettisoned him through one of the panes of a sliding glass door. There was broken glass everywhere. From then on, it was as if everything was happening in slow motion. I remember Tony looking at me with this stunned, sort of frozen expression as the blood poured from his little face. I remember rushing to Tony, scooping him up, and grabbing a clean terry-toweling nappy from a pile of clothes. My days as a St. John's Ambulance volunteer had taught me to check for shards of broken glass. When I was happy that there were none, I began pressing the nappy onto his face, applying the pressure as hard as I could to stem the flow of blood. I then cradled him in my arms and headed for Ellie, who was miraculously sitting still in the middle of this sea of broken glass. I scooped her up under my spare arm and sat there on my knees calling for help. All the while Purdey was running around like a lunatic, barking and jumping in the air as if she was playing some huge game.

It was every parent's nightmare. When help eventually arrived, friends and family were unanimous. Tony's injuries were awful and would leave him scarred for life. "This dog is a bad one, she's a rogue," they said. I still felt responsible for Purdey, however, and was determined to give the dog another chance. She continued getting herself into problems every now and again, but, for a couple of months at least, all was relatively calm.

Then one sunny winter's morning, just before Ellie's first birthday in February, I was in another part of the house while Ellie was on the floor playing with her toys, supervised by my mother. The moment I heard my mother scream, I realized something had happened. When I got to the sitting room, my mother just shouted, "The dog's bitten her. Ellie did nothing and the dog's bitten her. The dog's turned." I didn't want to believe it. But when I saw Ellie had a rather nasty little nick over her right eye, I had no option. My head was spinning. Why had this happened? What had Ellie done? Where had my training gone wrong? I knew, however, that the time for questions was over.

As soon as he heard the news, my father came around to see me. As a girl, I had heard him talk of one of his favorite dogs, an Old English sheepdog cross called Gyp, and how he had "turned." My grandmother had been trying to move him off a sofa and he had snapped at her. In my grandfather's mind if a dog could turn on the hand that fed it, then it was doomed. So Gyp was destroyed. My father did not have to spell it out for me. "You know what you've got to do, my girl, once they've gone, they've gone," he said sadly. "Don't waste your time, just do it." That evening the children's father came back from work. "Where's the dog?" he asked me. "She's dead," I told him. I had taken her to the vet that afternoon and had her put down.

For a long time, part of me believed I had done the right thing with Purdey. Yet at the same time I always felt that I failed her, that it was my fault not hers. Even when I had her put down, I felt I was deserting her. It took me almost twenty years to confirm my suspicions. What I now know is that Purdey's behavior was all caused by my inability to understand that dog, to communicate with her, to show her what I actually wanted. In the most simple terms: she was a dog, a member of the canine not the human family, yet I was using a human language.

Over the past ten years, I have learned to listen to and understand canine language. As that understanding has grown, I have been able to communicate with dogs, to help them—and their owners—overcome their problems. On many occasions, my intervention has prevented a dog from being destroyed because of its seemingly untreatable behavior. The joy I have felt each time I have saved a dog's life in this way has been immense. I would be lying if I did not admit that it is also tinged with regret that I did not learn these principles in time to save Purdey.

The object of this book is to pass on the knowledge I have acquired. I will explain how I arrived at the method I now operate. I will then go on to outline how you can learn this language for yourself. Like all languages, it has to be treated seriously. Learn it lazily or half-heartedly and it will only confuse both you and the dog with which you are trying to communicate. Learn it well and I can assure you that your animal will reward you with cooperation, loyalty, and love.

I

The Lost Language

"THE DOG IS A LION IN HIS OWN HOUSE."

Persian Proverb

Mankind has misplaced many secrets in the course of its history. The true nature of our relationship with the dog is among them. Like many millions of people around the world, I have always felt a special affinity exists between our two species. It goes beyond mere admiration for the dog's athleticism, intelligence, and looks. There is an intangible bond there, something special that connects us and probably has done since our earliest beginnings.

For most of my life, this feeling was founded on little more than instinct, an act of faith, if you like. Today, however, the subject of man's relationship with the dog is the subject of a burgeoning body of intriguing scientific evidence. That evidence indicates that the dog is not only man's best friend but also his oldest.

According to the most up-to-date research I have read, the two species' stories became intertwined as long ago as 100,000 years BC. It was then that the modern human, *Homo sapiens,* emerged from his Neanderthal ancestor in Africa and the

Middle East. It was also around this time that the dog, *Canis familiaris,* began to evolve from its ancestor, the wolf, *Canis lupus.* There seems little doubt that the two events were connected and that the link lies in man's earliest attempts at domestication. Of course, our ancestors have incorporated other animals into their communities, most notably the cow, the sheep, the pig, and the goat. The dog, however, was not just the first but by far the most successful addition to our extended family.

There is compelling evidence to suggest our forefathers valued their dogs above almost everything else in their life. One of the most moving things I have seen in recent years was a documentary on the discoveries made at the ancient Natufian site of Ein Mallah in northern Israel. There, in this parched and lifeless landscape, the 12,000-year-old bones of a young dog were found resting beneath the left hand of a human skeleton of the same age. The two had been buried together. The clear impression is that the man had wanted his dog to share his last resting place with him. Similar discoveries, dating back to 8500 BC, have been made in America, at the Koster site in Illinois.

The sense that man and dog had a unique closeness is only underlined by the work done by sociologists in communities in Peru and Paraguay. There, even today, when a puppy becomes orphaned, it is common for a woman to take over the rearing process. The dog feeds off the woman until it is ready to stand on its own feet. No one can be sure how far back this tradition goes. We can only begin to guess at the intensity of the relationship these people's ancestors must have had with their dogs.

There are, I'm sure, many more discoveries to be made, many more eye-opening insights to be gained. Yet even with the knowledge we now have, we should not be surprised that the empathy between the two species was so powerful. Quite the opposite in fact; the immense similarities between the two animals made them natural partners.

The wealth of study that has been done in this area tells us that both the ancient wolf and the Stone-Age man shared the same driving instincts and the same social organization. In simple terms, both were predators and lived in groups or packs with a clear structure. One of the strongest similarities the two shared was their inherent selfishness. A dog's response to any situation—like man's—is "what's in it for me?" In this instance, it is easy to see that the relationship they developed was of immense mutual benefit to both species.

As the less suspicious, more trusting wolf settled into its new environment alongside man, it found it had access to more sophisticated hunting techniques and tools such as snares and stone arrows, for instance. At night it could find warmth at the side of man's fire and food in the form of discarded scraps. It was little wonder it took so easily to the domestication that was about to begin. By introducing the wolf to his domestic life, man reaped the benefits of a superior set of instincts. Earlier in his history, the Neanderthal man's exaggerated proboscis had provided him with a powerful sense of smell; his descendant saw that by integrating the newly domesticated wolf into the hunt, he could once more tap into this lost sense. The dog became a vital cog in the hunting machine, helping to flush out, isolate, and, if necessary, kill the prey. In addition to all this, of course, man enjoyed the companionship and protection the dog provided within the camp.

The two species understood each other instinctively and completely. In their separate packs, both man and wolf knew their survival depended on the survival of their community. Everyone within that community had a role to perform and got on with it. It was only natural that the same rules should be applied in the extended pack. So while humans concentrated on jobs like fuel gathering, berry picking, house repairs, and cooking, the dogs' main role was to go out with the hunters as their eyes and ears. They would perform a similar role back within the camp, acting as the first line of defense, warding off attackers and warning the humans of their approach. The degree of understanding between man and dog was at its peak. In the centuries that have passed since then, however, the bond has been broken.

It is not hard to see how the two species have gone their separate ways. In the centuries since man has become the dominant force on earth, he has molded the dog—and many other animals—according to the rules of his society alone. It did not take man long to spot he could adjust, improve, and specialize the skills of dogs by putting them together selectively for breeding purposes. As early as 7000 BC, in the Fertile Crescent of Mesopotamia, for instance, someone noticed the impressive hunting skills of the Arabian desert wolf, a lighter, faster variety of its northern relative. Slowly the wolf evolved into a dog able to chase and catch prey in this harsh climate and, more importantly, to do so according to man's commands. The dog—variously known as the saluki, Persian greyhound, or gazellehound—remains unchanged today and may well be the first example of a purebred dog. It was certainly not the last. In ancient Egypt, the Pharaoh hound was bred for hunting. In Russia, the borzoi was bred to chase bears. In Polynesia and Central America, communities even developed dog breeds specifically for food.

The process has continued through the ages, aided by the dog's willingness to be "imprinted" by our species. Here in England, for instance, the hunting culture of the landowning aristocracy produced a collection of dogs customized to fulfill specific roles. On a nineteenth-century estate, a typical pack would include a springer spaniel to literally spring or flush the game from cover, a pointer or setter to locate birds, and a retriever to return the dead or wounded game to the handler.

Elsewhere, other breeds maintained the historic bond between man and dog even more closely. Nowhere was this exemplified better than in the development of guide dogs for the blind. It was at the end of the Great War, at a large country convalescent home in Potsdam, Germany, that a doctor working with injured veterans noticed just by chance that when patients who had lost their sight started moving toward a flight of steps his German shepherd would cut them off. The doctor sensed the dog was turning them away from danger. He began training his dogs specifically to use this natural shepherding ability to help humans who could no longer see. The guide dog for the blind developed from there. It may be our most direct throwback to that earliest community. Here was a dog providing a sense that man has lost. Unfortunately, it is a rare example of cooperation in the modern world.

In more recent times, our relationship has changed, as far as I am concerned often to the detriment of the dog. Our former partners in survival have become companions cum accessories. The evolution of the so-called lapdog illustrates this perfectly. The breeds were probably begun in the Buddhist temples of the high Himalayas. There, holy men bred the hardy Tibetan spaniels so that they became smaller and smaller. They then used the dogs as body warmers, teaching them to jump up onto their laps and remain under their robes to fend off the cold.

By the time of Charles II, the idea had traveled to England, where the English toy spaniel evolved from breedings of tinier and tinier examples of the setter. Over time, these little gun dogs were pampered by their wealthy owners and crossed with toy-dog breeds from the East. The breed's history is still visible today in the distinctive flat-faced features of the King Charles spaniel. This was, to my mind, a pivotal moment in the history of man's relationship with the dog. To the dog, nothing had changed, but to his former partner, the relationship was entirely new. The dog had ceased to have a function beyond mere decoration. It was a foretaste of what was to come.

Today, examples of the old relationship that man and dog

enjoyed are few and far between. Working dogs such as gun dogs, police dogs, and farm dogs, as well as the guide dogs I have already mentioned, spring to mind. However, they are the tiny exceptions. In general, today we have a culture and society in which no consideration has been given to the dog's place. The old allegiance has been forgotten. Our familiarity has bred contempt, and along the way the instinctive understanding the two species shared has been lost.

Again, it is easy to see why there has been a communications breakdown: the small communities in which we began our history have been replaced by one huge, homogeneous society, a global village. Our lives in the big cities have made us anonymous, and we don't know or acknowledge the people we are around. If we have become divorced from the needs of our fellow humans, we have lost touch completely with dogs. As we have learned to cope with all the things we have to face in our society, we have simply assumed that our dogs have done the same thing. The truth is they haven't. Today, man's concept of the dog's role and the dog's idea of its place are completely at odds with each other. We expect this one species to abide by our norms of behavior, to live by rules we would never impose on another animal, say a sheep or a cow. Even cats are allowed to scratch themselves. Only dogs are told they cannot do what they like.

It is ironic—and to my mind, tragic—that of all the 1.5 million species on this planet, the one species blessed with the intelligence to appreciate the beauty in others fails to respect dogs for what they are. As a result, the exceptional understanding that existed between us and our former best friends has all but disappeared. It is little wonder there are more problems with dogs today than there have ever been.

Of course, there are many people who are living perfectly happily with their dogs. The ancient bond clearly lives on inside us somewhere. No other animal evokes the same set of emotions or forms the basis for such loving relationships. The fact remains that people today who are living in harmony with their dogs are getting there by a happy accident rather than through knowledge. Our awareness of the instinctive, unspoken language that we share with our dogs has been lost.

In the last decade, I have attempted to bridge that divide, to try to re-establish that link between man and dog. My search for this missing means of communication has been a long and at times frustrating one. Ultimately, however, it has been the most rewarding and exciting journey I have ever made.

2

A Life with Dogs

It is hard for me to imagine this now, but there was a time when I could not face the prospect of forming a friendship with another dog. In the awful aftermath of Purdey's death, I had become deeply disillusioned. At one point I even think I came out with the classic line "I will never have another dog in this house." The reality was, however, that my affection for dogs ran too deep. And, within a year or so of Purdey's death, a little gun dog was healing the scars left by my tragic loss.

Despite our early setback, my family and I had settled well into country life. It was my husband's interest in hunting that brought dogs back into our home. One day in the autumn of 1973, he came back from a rough shoot bemoaning his lack of a good gun dog. He had seen a wounded rabbit slinking its way into the woods to die. "If I had a dog that couldn't have happened," he complained with a look that left little room for doubt about what he was thinking.

So it was that on his birthday that September, his first gun dog, a springer spaniel bitch we called Kelpie, arrived in the house. He loved the dog as I did. It was the beginning of my lifelong love affair with that beautiful breed.

We were, predictably I suppose, terrified of repeating the experience of Purdey and immediately bought one of the standard text books on gun dog training. I have to confess that our first efforts at shaping Kelpie up were far from a roaring success. We wanted to train Kelpie to retrieve, an unnatural act for a springer. Sticking rigidly to the book, we started her off by throwing objects for her to recover and return to us. The book stressed the importance of beginning with something very lightweight. The idea was to teach the dog to be "soft mouthed" with the objects it recovered.

We decided to use one of Ellie's old bibs, which we tied in a knot. One morning we took Kelpie outdoors, threw the bib into the distance, and waited for her to return it to us. We were so thrilled when she bounded off and picked up the bib, but our expressions soon changed as she ran straight past us into the house. I remember my husband looking at me with a blank look. "What does the book say we do now?" he asked. At that point, I think we all collapsed to the floor with laughter. We made an awful lot of mistakes with Kelpie, but we had great fun too. Whenever I feel too full of myself or over-confident about the control I am able to achieve over dogs today, I think back to that moment.

Kelpie was very much my husband's dog, however. I was so pleased with her and the way she had fitted in so well to our life that soon afterward I decided to get a dog of my own. I had fallen hopelessly for the spaniel and bought a nine-week-old puppy, a bitch from the show strain of the springer spaniel. I called her Lady after the imaginary dog I'd had as a child.

My interest lay less in hunting than in breeding and showing dogs. So it was that Lady became my introduction to that fascinating world. By the middle of the 1970s, I was traveling with her to shows all over the country. She was a lovely dog and was popular with judges wherever we went. By 1976, Lady had qualified for the most prestigious dog show of all, Cruft's, in London. The day we traveled down to the famous arena at Olympia was a moment of great pride for me.

I found the world of dog shows rewarding and hugely enjoyable. It was, apart from everything else, a great social network, a way of meeting like-minded people. Two of the closest friends I made were Bert and Gwen Green, a well-known couple in the dog world, whose line of dogs, under the Springfayre affix, were hugely popular. Bert and Gwen knew of my interest in moving on to breeding dogs. It was they who gave me Donna, Lady's three-year-old grandmother. Donna had all the makings of a good, foundation bitch and helped me start my own breeding line. I had soon bred my first ever litter from her, and kept one of the seven dogs for myself, calling him Chrissy.

Chrissy was a show dog that became a very successful working gun dog. He won a puppy class at the age of eight months and qualified for Cruft's too. The highlight of my time with him came in October 1977 when I took him to the Show Spaniels Field Day, a prestigious event for gun dogs that have qualified for Cruft's. The competition judged the dogs on their working ability only. I was, as the footballing expression goes, over the moon when Chrissy won the prize for Best English Springer On The Day. I vividly remember the moment the judge handed me the winner's rosette. "Welcome to the elite," he told me. After that, I truly felt I had arrived in the dog world.

Encouraged by this success, I went on to improve my line through two well-bred bitches, and I think I gained a pretty respectable reputation. Throughout this time, I was also adding to the family's collection of dogs. Tragically, Donna died of a tumor in 1979, aged only eight; but in the aftermath, I also bought a cocker spaniel named Susie for my daughter and bred from her daughter Sandy.

It was, however, Khan, one of the English springer spaniels I had bred, that brought me my greatest success, winning many classes and Best of Breed. He was a wonderful dog with beautiful features, in particular the sort of warm but masculine face that judges were always looking for. In 1983, he qualified for Cruft's, emulating the feat of six of my previous dogs. To my delight he won his class. Again, the memory of receiving the winner's card fills me with pride.

As I have explained, I met some wonderful, warm-hearted people who taught me a great deal. There was no wiser soul than Bert Green. I remember he used to say to me: "I doubt you do the breed any good, but don't do it any harm." By that he meant we had a responsibility to be faithful to the principles of the dog-breeding fraternity.

To me, breeding dogs came with its own set of responsibilities, particularly as the majority of the small number of dogs I bred were being carefully placed into family homes. My job was to ensure these dogs had temperaments that made them a pleasure to own. So inevitably I had spent a lot of time working on training the dogs, working on what everyone generally referred to as "obedience classes."

It was here that the unease I had long felt about our attitude to dogs really broke through to the surface. The memory of Purdey was a constant cloud at the back of my mind. I was forever asking myself what I had done wrong, wondering whether I had somehow given her the wrong kind of training?

My growing unease was fueled further by the mistrust I felt about the traditional enforcement methods of training. There was nothing radical or revolutionary about my training techniques then. Far from it, I was as conservative as everyone else in most ways. I would go through the routine of teaching a dog to sit and stay by pushing its bottom on the ground, to come to heel with a jerk on a choke chain, and to follow. And I would instill these disciplines through the time-honored methods.

Yet as I spent more and more time training, I became aware of a nagging doubt about what I was doing. It was as if a voice at the back of my mind was constantly saying: you are making the dog do this, the dog does not want to do this.

In truth, I had always hated the word "obedience." It carried the same connotation as "breaking in" within the horse world. It simply underlined the reality of the situation, that what I was using was a kind of enforcement, a means of going against the will of the animal. It is, to my mind, like the word "obey" within marriage vows. Why not use words like "work alongside," "pull together," "cooperate"? "Obey" is just too emotive for me. But what could I do about it? There were no books about how to do it any other way. And who was I to argue? There are no two ways about it. You have to have your dog under control. You can't have it just running amok. It is our responsibility as it is with our children to make them socially responsible. I had no real alternative.

Nevertheless, it was at this time that I began trying to make the training process more humane if I possibly could. With this in mind, I began introducing a few subtle changes in my technique. The first involved nothing more complex than a simple change of language. As I explained, I was using the traditional methods of enforcement, including the so-called choke chain. As far as I was concerned, the name was a misnomer. Used correctly, the chain should never choke a dog. It should merely check it. There was no use in using it to jerk dogs back as far as I was concerned. So I tried to soften the terminology so as to soften the attitude of the humans.

In my training, I taught people to use the chain to make a light, clicking noise that the dog would recognize as an anticipatory signal before it moved forward. When it heard the chain, it reacted so as to avoid being choked. So to me and my pupils, they were check chains rather than choke chains. It was a minor change, but the difference in emphasis was fundamental.

I tried to do the same in heel work. I did not approve of the method most people used which involved taking the lead and pulling the dog down. I thought that was wrong. My original way of getting it to lie down was to make the dog sit, then tip the dog gently to one side by taking away its inside leg. Wherever I could, I was always looking for a softer way within the traditional parameters of the work.

As I did so, I was very successful at teaching people how to work with their dogs. Yet the changes I was achieving in softening the approach were so small. The central philosophy remained the same. I was making the dog do it. I always felt I was imposing my will on the dog rather than making it do what I wanted by choice. And I sensed that the dog did not know why it was doing it. The ideas that changed all this began to form themselves at the end of the 1980s.

By that time, my life had changed considerably. I had been divorced and my children were growing up and on the road to university. I myself had studied psychology and behaviorism as part of a degree in literature and social sciences at Humberside University. I had to give up showing dogs because of the divorce. Just as people were beginning to respect me and I was beginning to knock on the door, it was all kicked away; it was very frustrating. I reluctantly had to let some of my dogs go.

Meanwhile, I maintained a pack of six dogs. By the time we moved to a new home in North Lincolnshire in 1984, there was little time for life in the competitive dog world. I was working too hard to support my kids to be able to afford to compete or to breed full time. Apart from my own dogs, my contact with that world was confined to working at the local Jay Gee Animal Sanctuary and writing a pet page for a local newspaper.

My passion for dogs remained as great as ever. The only difference now was that it had to be channeled in a different direction. My interest in psychology and behaviorism had carried on from university. Behaviorism in particular had really become part of the mainstream by now. I had read Pavlov and

Freud, B.F. Skinner, and all the acknowledged experts in the field and, to be honest, I found a lot that I could agree with. The idea, for instance, that when a dog is jumping up, it is aiming to establish a hierarchy, and is jumping so as to put you in your place. Or the idea that a dog will barge its way in front of you as you walk to a door because it is checking the coast is clear, protecting the den, and believes it is the leader.

I also understood and accepted the idea of what was referred to as "separation anxiety." The behaviorists' view was that a dog will chew up the furniture or destroy the home because it is separated from its owner and that separation is stressful for the dog. All these things made total sense and offered me a lot. But to me there was something missing. What I kept asking was: why? Where was the dog getting this information from? At the time I wondered whether I was crazy for even asking myself this, but why is a dog so dependent on its owner that it is stressful to be separated? I didn't know it then, but I was looking at the situation the wrong way around.

It is not an understatement to say that my attitude to dogs—and my life—changed one afternoon in 1990. By this time, I was also working with horses. The previous year, a friend of mine, Wendy Broughton, whose former racehorse, China, I had been riding for some time, had asked me if I was interested in going to see an American cowboy called Monty Roberts. He had been brought over by the Queen to demonstrate his pioneering techniques with horses. Wendy had watched him give a demonstration in which he had brought a previously unsaddled horse to carry saddle, bridle, and rider within thirty minutes. It was, on the surface at least, highly impressive, but she remained skeptical. "He must have worked with the horse before," she thought. She was convinced it had been a fluke.

In 1990, however, Wendy had been given the chance to put her mind at rest. She had answered an advert Monty Roberts had placed in *Horse & Hound* magazine. He was organizing another public demonstration and was asking for two-year-old horses that had never been saddled or ridden before. He had accepted Wendy's offer to apply his method to her chestnut thoroughbred mare, Ginger Rogers. In truth, Wendy saw it as a challenge rather than an offer. Ginger Rogers was an amazingly headstrong horse. Privately we were convinced Monty Roberts was about to meet his match.

As I traveled to the Wood Green animal sanctuary near St. Ives, Cambridgeshire, on a sunny, summer's afternoon, I tried to keep an open mind, not least because I have immense respect for the Queen's knowledge of animals, her horses and dogs in particular. I thought if she was giving credence to this fellow then he had to be worth watching.

I suppose when you hear the word "cowboy," you immediately conjure up images of John Wayne, larger-than-life characters in Stetsons and leather chaps, spitting and cursing their way through life. The figure that emerged before the small audience that day could not have been further removed from that cliché. Dressed in a jockey's flat cap, wearing a neat, navy shirt and beige slacks, he looked more like a country gentleman. And there was nothing brash or loud about him. In fact, he was very quiet and self-effacing. But there was undoubtedly something charismatic and unusual about him. Just how unusual, I would soon find out.

There were about fifty of us sitting around the round pen he had set up in the equestrian area. Monty began by making some opening remarks about his method and what he was about to show. The early portents were not good, however. Unknown to Monty, Ginger Rogers was behind him. As he spoke, she started nodding her head slowly, almost sarcastically pretending to agree with him. Everyone burst out laughing.

Of course, when Monty turned around, Ginger stopped. The minute he swiveled around to face the audience again, she started again. Wendy and I looked at each other knowingly.

We were both thinking the same thing I'm sure: he's taken on too much here. As Monty gathered up a sash and began going through the opening of his routine, we sat back waiting for the fireworks to begin.

Precisely twenty-three-and-a-half minutes later, we were ready to eat our words. That was how long it took Monty not just to calm Ginger down, but also to have a rider controlling with ease a horse that to our certain knowledge had never been saddled or ridden in its life. Wendy and I sat there in stunned silence. Anyone who saw us that day would have seen disbelief written all over our faces. We remained in a state of shock for a long time afterward. We talked about it for days and days. Wendy, who had spoken to Monty after his miraculous display, even went on to build a replica of his trademark round pen and started implementing his advice.

For me too it was as if a light had been switched on. There were so many things that struck a chord. Monty's technique, as the whole world now knows, is to connect—to "join up" in his phrase—with the horse. His time in the round pen is spent establishing a rapport with the horse, in effect communicating in its own language. His method is based on a lifetime working with and, most importantly, observing the animal in its natural environment. Most impressive of all, his method has no place for pain or fear. His view was that if you did not get the animal on your side, then anything you did was an act of violation. You were imposing your will on an unwilling being. And the fact that he was succeeding in doing things differently was clear from the way he won the trust of the horse. He placed great store, for instance, on the fact that he could touch the horse on its most vulnerable area, its flanks. That day, as I watched him working in unison with the animal, looking at and listening to what the animal was signaling to him, I thought "he's cracked it." He had connected with the horse to such an extent that it let him do whatever he liked. And there was no enforcement, no violence, no pressure: the horse was doing it of its own free will. I thought how the heck can I do this with dogs? I was convinced it must be possible given that dogs are fellow hunter-gatherers with whom we have a much greater connection historically. The big question was: HOW?

3

Listening and Learning

I realize now that fortune was smiling on me at this time. If I had not begun expanding my own pack of dogs, I am sure I would never have seen what I did. By this time, my pack was reduced to a quartet of dogs: Khan, Susie, and Sandy, and a beagle I had taken in, called Kim. They were a fun foursome, a wonderful mixture of characters. By now, however, I was entering another new phase in my life. I had no ties, the kids were grown up, and I had just lost my parents. Free to think about what I wanted to do, I decided to welcome a beautiful, black German shepherd puppy called Sasha into my home.

I had always liked the idea of owning a German shepherd even though they're a breed that has had a bad press. People see them as police dogs, aggressive animals that are always attacking people, which is, of course, far removed from the truth. We stereotype dogs just the same way as we pigeonhole people. All German shepherds are aggressive, all spaniels are stupid, or all beagles are wanderers—we have all heard it. Yet it is just as ignorant as saying all Frenchmen wear berets or all Mexicans walk around in sombreros—it is nonsense. My reluctance to take on a German shepherd was nothing to do with this. I quite simply didn't think I was good enough to work with this kind of dog. I had heard a lot about their immense intelligence, about how you have to challenge their brains, give them something to think about. I always felt I didn't have the time, the patience, and certainly the knowledge to handle one. Now, perhaps, I did.

Sasha's arrival in my home marked a major turning point. After watching Monty in action, I knew that I had to follow his example and observe very closely what my dogs were doing. I had to stop thinking I knew best and start watching them. As I did so, the benefits were not long in coming. Sasha was a young and incredibly energetic dog. My other dogs reacted to this exuberant new presence in different ways. The beagle, Kim, would simply ignore her. Khan, on the other hand, was quite content playing with the newcomer. He did not mind at all that Sasha would follow him everywhere, sticking like glue day and night. It was Sandy, my son Tony's cocker spaniel, who had the problems.

From the moment Sasha arrived in the house, Sandy made it plain that she hated this newcomer. Sandy to be fair was getting on. She was twelve years old by now, and she simply didn't want this energetic young kid leaping all over her. At first she tried ignoring her by turning her head from side to side, which was sometimes difficult because Sasha at the age of ten weeks was bigger than Sandy. When this didn't work she began making this low grizzling sound and curling her lip so that Sasha would back off.

As I sat down and wondered about what was going on here, I realized it was something I had seen before in another dog of mine, one of my original springer spaniels, Donna, or The Duchess as she became known. As her name suggests, there was something regal about Donna. When she moved around the house, everyone had to move out of the way. I remember on one occasion my mother arrived and sat down in the armchair which Donna used. Donna had been lying there quite happily curled up. The moment my mother sat next to her, she lifted herself up, looked up indignantly, and pushed her off the edge. My mother ended up on the floor. When she got up and sat there again, the same thing happened. Donna pushed her off again. At the time, of course, we thought it was hilarious.

As I watched Sasha and Sandy, I realized I was seeing something similar happening again. I had seen this in the past without realizing what I was watching. Now, however, it was as if I was witnessing things for the first time. It was fairly obvious what was going on here: Sandy like Donna was trying to show who was the boss. It was to do with a status of some kind.

The next thing I noticed was the very intense performance my dogs would go through whenever they came together. If, for instance, I took Sasha to the vet for an injection, each time she came home, she would immediately go through this performance. I didn't know what to call it at the time but now I would say it was a ritualized greeting. There would be a lot of her licking all the other dogs' faces with her ears pinned back: it always happened.

At first it didn't make much sense to me. In Sasha's case I didn't know whether to put it down to youthful exuberance, her newness to the group, or some habitual thing she had picked up before arriving in my home. Luckily, Sasha's inspiration was not confined to her actions. In her looks she reminded me very much of a wolf. I had read a little about wolf packs in the past but she made me think about it more closely.

I got out some videos on wolves, dingoes, and wild dogs and was amazed when I immediately saw this same sort of behavior. I was fascinated to see that in situation after situation they too went through this same ritualized greeting. I was fairly sure this was something to do with social status. That hunch solidified as I looked further into the mechanics of the wolf pack, a community in which everything revolves around the leaders—or the Alpha pair.

I will look at the Alpha pair in more detail later. For now I will simply explain that the two Alpha wolves are the strongest, healthiest, most intelligent, and most experienced members of the pack. Their status is maintained by the fact that they are the only members of the pack who breed, thus ensuring only the healthiest genes survive. The key point here is that the Alpha pair dominate and dictate every aspect of pack life. The remainder of the pack accept the Alpha pair's rule and defer to them without question. Below the leading pair, each subordinate member is content to know its vital place and function within this pecking order.

Watching films on wolves, it was obvious that the ritual greetings I was watching were all related to the wolves that seemed to be the Alpha pair. The wolves who seemed to be in charge did not lick the faces of the other wolves—the others all licked their faces. This licking was very specific in nature too; it was almost frantic and concentrated on the face. There were other clues in the body language too. The Alphas had a confidence level, an attitude. They carried themselves differently physically—most noticeably they carried their tails much higher than the others. The subordinates sent out their signals too. Some would simply drop their bodies below their leaders. Some wolves, presumably the younger and lower-ranking subordinates, would not even come that far forward; they would hang back. It was as if only some of the dogs were entitled to lick the leader, some of them were not.

Again, I quickly realized I had seen this before. The Duchess, my dog Donna, had carried herself in exactly the same imperious way. But it was when I went back to my pack at the time that the similarities really struck home. I immediately began seeing the same thing again. I saw it was as though there were kings, knights, and servants. It was clear that the lower-order dogs were being put in their place by those above them, just as within the wolf pack. I had never made the connection before.

Suddenly, I saw that dogs were the same. It was a huge step forward for me.

Again, it was Sasha who provided the most powerful proof. It was clear to me by now, for instance, that she had acquired a higher status within the pack. She had grown enough in size and confidence to ignore Sandy's protestations. Sandy at the same time had become more resigned about matters. She would tip her head away, dip her carriage and her tail.

The power shift was most obvious at playtime. When I threw the ball or whatever toy we were using, it would be Sasha's job to recover it. The others would follow it and bound around it when it landed, but there was no argument over whose role it was to retrieve the ball. And if another dog came near her once she had picked it up, Sasha would give them a little look, her whole body language would shout: "It's mine, now back off."

Sandy's body language in comparison was submissive, her body dropped lower and lower as this interaction went on. Sandy had in effect given up the fight and allowed Sasha to impose herself as the head of the pack. The younger dog had, if you like, staged a bloodless coup.

Of course, my dogs were not always displaying such intriguing behavior. There were times when they were happy in their own company. I began to understand that this hierarchy was being reinforced at particular times only. So the next step was to work out exactly when this communication was going on.

I noticed that this would happen with me whenever I got home. Watching the dogs more closely, however, I saw that the same sort of behavior was repeated with me whenever somebody else came to the front door. As the visitor came in, the dogs would crowd around me. They would get very excited, rush to the door, rush around the person. All the time they were doing this, they would be interacting, repeating this ritualized behavior. I saw the same thing again when I got the dogs' leads out and we got ready for a walk. All the dogs would get excited and agitated, jumping up and down and again interacting with each other as we got ready to leave the house.

Once more, I studied the wolf pack and once more I saw the same thing. In the wolves' case this behavior was occurring as the pack got ready to go on a hunt. There was a lot of running around and jostling for position, but ultimately it was the Alpha pair whose heads remained erect and their tail carriage high. And it was always they who led the pack away in search of the prey.

I realized the wolves were re-establishing who was in charge here. The leader was reminding the rest that it was his role to lead and theirs to follow. This was the pecking order and they must abide by it to survive. Clearly my pack was doing the same. What really interested me at this point, however, was my inclusion in all this. From the way my dogs were reacting around me, it was clear that I was somehow part of this process. And of all my dogs, none was so keen to involve me in the process as Sasha.

If we were going out of the house, Sasha would invariably stand in front of me. She would place herself in a position, across my body, blocking me off. Although I could hold her back with my chain, she always wanted to go ahead of me. She seemed to think it was natural for her to go forward first. Equally, if there was a loud noise or an unexpected event while we were out on a walk, such as another dog appearing in front of us, she would stand in front of me in a very protective stance. She would also bark more furiously than the others when someone went past the house in view or when the postman or milkman came to the door. And unlike the others, there seemed to be no calming her in these situations.

If I am honest about it, part of me was worried about this behavior. It reminded me a little of Purdey who also had this habit of running around in front of me. For a while, part of me feared I might let my dog down again. Fortunately this time, however, I saw what was going on. Again, memories of Donna provided a first clue. I recalled how she had behaved years earlier when I had fostered a little baby boy, Shaun. Whenever he lay on his blanket on the floor, Donna would lie next to him with her leg over his leg. If he kicked it off, she would move it back. She was clearly acting as his protector, guarding over him at all times. It was now that I realized that, just as Donna had felt the little baby was her responsibility, somehow Sasha must also be feeling she had a role to perform in looking after me. Why else would I be given such specific treatment when I came in through the door or when greeting visitors? Why else would she get so hyperactive about my leading her out on a walk?

I realize now that so many of my mistakes were down to human conditioning. Like almost every other person on this planet, I had assumed that the world revolved around our particular species, and that every other species had somehow fitted into our grand scheme. I had assumed that because I owned the dogs, then I had to be their leader too. Now, for the first time, I began to wonder if that really was the case. I began to wonder whether Sasha was trying to take care of me.

All of the information I was getting from my dogs was powerful. But this to me was the most explosive knowledge of all. It made me completely re-evaluate my thinking. And it was then that the penny began to drop. I thought: "Hold on, what if I am looking at this the wrong way around? What if I am imposing a rather arrogant, presumptuous—and typically human—framework on this? What if, instead, I imagine it from a dog's point of view and rather than thinking it is dependent on us, the dog thinks the exact opposite, that it is responsible for us? What if it believes it is the leader of a pack in which we too are subordinates? What if it believes it is its job to safeguard our welfare rather than the other way around?" As I thought about it, so much suddenly locked into place.

I thought of separation anxiety. Instead of looking at a dog that was worrying "Where's my mum or dad?" we had a dog worrying "Where's my damned kids?" If you had a two-year-old and realized you didn't know where it was, wouldn't you be going insane with worry? Dogs were not destroying the house through boredom; it was through sheer panic. When your dog jumps up at you when you come in, it is not because it wants to play with you, it is because it is welcoming you back to the pack that it believes it is in charge of.

In many ways I felt a fool. I had made the mistake we humans make all too often in our dealings with animals. I had assumed my dogs did not have their own language. How could they—they lived with us? I had assumed that they understood they were living with me in a domestic situation. It had not occurred to me to think the rules they were playing by had been dictated to them in the wild. In short, I had imposed human constraints on them; I had allowed familiarity to breed contempt. I can't say the idea came to me in one blinding flash. There was no apple falling from a tree or a bolt of lightning in the sky. But from that moment, my entire philosophy changed.

4

Taking the Lead

In a few short months I had gained a greater insight than I would have imagined possible. By taking time to watch my dogs interacting with each other, by listening to what they were telling me, I had picked up on some powerful knowledge. Behavior I had seen in the wild was being repeated on a daily basis in my own home by my own dogs. I had begun to see how they enforced their will on others, how they showed supremacy, how they showed dominance. And there was no shouting because dogs don't shout, no smacking because dogs don't hit.

From my dogs, I had isolated three clear occasions when interaction was going on between them: at times of perceived danger, when they were going for a walk, and when they were reuniting. At each of these times, I saw certain dogs being put in their place, the leader asserting its authority and the subordinates accepting that authority. What I wanted to know now was, how could I take this a step further?

To my mind, the most inspirational aspect of Monty Roberts's work was the way he was able to replicate the behavior of a horse even though he was a human. I knew that I had to try to follow his example and reproduce the behavior of my dogs. I wanted to see how much difference it would make if I took charge in the way that a leader would do in the wild. I also, crucially, wanted to find out if it was something that should be done. Would there be any side effects, how would it impinge on the dogs' well-being and quality of life? With this in mind, I knew the most important challenge was to develop a way of leading the dogs to decisions they were making of their own free will. As Monty puts it, I wanted a situation where if there was a meeting, I would be elected chairman. It was a daunting task.

Before I started, I knew two elements were of paramount importance. I was soon calling them "the two Cs." I had to be consistent and I also had to be calm. For generations we have been taught to instill obedience in our dogs by barking orders at them. Words like "sit, stay, beg, come," we have all used them. I use them myself. Dogs do recognize them, but not because they understand the meaning of the words. They merely learn to make associations with the sounds if they are used repeatedly. As far as I am concerned, their effectiveness proves only the value of being consistent in providing information to your dog. In every other respect, shouting at the top of your voice is a surefire way of creating a neurotic dog.

As I got ready to take the next step, this feeling was reinforced all around me. In the park where I used to exercise my dogs, I remember a man who used to exercise his Doberman. Any dog approaching the Doberman was greeted with the owner shouting and shaking a walking stick. Almost as soon as he started doing this, his dog would start growling and snapping too. I noticed that, in contrast, people who were relaxed and happy with their dogs tended to be in charge of animals who were relaxed and happy at play. This got me thinking about the nature of the leadership I should be providing, and I quickly saw that calmness seemed to be a fundamental requirement for all sorts of reasons.

In both the human and the dog world, the greatest form of leadership is the silent, inspirational type. Think of the great men of history: Gandhi, Sitting Bull, Mandela—all hugely charismatic but quiet men. That famous phrase from Kipling's poem "If" always comes to mind when I think of the qualities of leadership:

If you can keep your head when all about you
Are losing theirs . . .

It is obvious when you think about it. A leader that is upset or agitated is a leader that does not instill confidence, a leader that is less likely to be believed in. It is certainly a principle that is recognized within wolf packs where the Alpha wolves display a serenity that borders on the dismissive at times.

I knew if I was going to begin communicating in my dogs' language and, more importantly, if I was going to be elected leader, I had to start behaving in a manner the dogs would associate with leadership. I am not by nature the strong, silent type, so it was necessary for me to adopt a slight change of personality in the dogs' company. Compared to the transformation I was soon seeing, the change was minor.

My first attempts began on a wet weekday morning. I remember it was raining really hard, and thinking how easy it would be to wait for a sunny day to start this bright new beginning. But by now I was impatient to get on. And I had gone to bed the night before, determined to try something the next day. I must admit I was full of self-doubt. I had no idea if it was going to work. Part of me felt a bit silly. I thought to myself: "I hope no one comes around this morning." But as I came down the stairs I knew that I had nothing to lose.

People imagine that I have always had my dogs behaving exactly as I wanted. They couldn't be further from the truth. At that time, my pack was quite a handful, and even worse, they had no manners. When I came home, they bounded around and jumped up just like any other dogs; it could be incredibly irritating. Sometimes I would have my arms full of shopping or I would be wearing a nice outfit and they would come careering at me. For this reason, the first situation I decided to tackle was the re-formation of the pack.

Planning it all in my mind the night before I began, I had decided to imitate the behavior of the Alpha by ignoring them. This was not, of course, the easiest thing in the world to do. But I soon realized that I had more tools available than I had thought. Because we are verbal creatures, we use words too much. We forget that we know an awful lot of body language as well. If somebody turns away from you, for instance, you know what they mean. Equally, if you walk into a crowded room and someone averts their eyes, you are getting a clear message straight away. Dogs use this same language too, eye contact in particular. I soon realized I could use this effectively. So when I came downstairs that morning and let the dogs into the kitchen, I started behaving differently. When they jumped up at me, I didn't say get down. When they misbehaved, I didn't tell them to go to their beds. For the first few minutes that day, I made sure I didn't even make any eye contact with them. I just ignored them.

It was, I confess, an unnatural feeling at first. I was cutting against an ingrained attitude that wanted to interact with dogs whenever possible. I'm not sure how long I would have been able to keep it up if I had not got almost immediate results. The impact was obvious within a day or two of my starting this new regime. To my astonishment, they very quickly stopped jumping up and charging at me. As I repeated the procedure each time I arrived among them, they became more and more respectful. As the week wore on, they began standing back and letting me come in unmolested.

I'm sure their acceptance was increased by the fact that there were immediate benefits to this. By giving me the body space I needed, they saw a distinct change in the atmosphere during the times I was with them: I was pleased to see them. The dogs learned that when I wanted to spend time, it was quality time. Behaviorism had taught me that you should ignore undesirable and excessive behavior but be sure to praise the positive, so I underlined this by making a quiet, extra fuss of them when they did come to me. The dogs were soon coming to me only when I asked them to. And it didn't take time; it happened within a week.

This first tentative step had proved so effective, I knew I was onto something. But I quickly realized that one thing alone was not going to give them the message. I decided to move on next to moments of perceived danger, and the arrival of strangers to the pack specifically. Like other dogs, mine used to bark incessantly when someone came to the door. When I let them in, the visitor would instantly be surrounded by a circle of dogs, jumping up at them and making a terrible fuss. I would shout: "Stop it, be quiet." But by now I realized that far from placating them, I was exacerbating the situation. Again I thought of Kipling; I knew I had to keep my head, be calm and consistent.

This time I decided to tell people to ignore the dogs when they came through the door. Those dogs that kept bounding up, I took into another room. Of course, some people thought I was crazy. To them, it was the most natural thing in the world to acknowledge a dog, particularly if it's a beautiful dog. My friends and family had certainly been in the habit of making a fuss of Sasha, Khan, Sandy, and Kim. But I was determined to give this a chance and insisted they do as I ask.

The early signs were enough to convince me to stick at it. Within a few days again, things began to calm down. Soon the dogs were just barking rather than running up and milling around visitors. Once more, the dogs picked up on what was being asked of them pretty quickly. Of course, I couldn't quite believe it was so simple; I put some of it down to the fact that both Sandy and Khan were getting old. I was sure there was significance in the fact that the dog that was giving me most in terms of response was Sasha, the youngest one in the pack and a German shepherd to boot. I never thought: "I'm right here, there has to be reasons why this is working"—I was questioning things all the way along. Despite all this, however, I can't deny it was a fantastic feeling. They were transformed. They seemed happier, calmer dogs, and it was a joy to behold.

The next thing I wanted to tackle was going for a walk. Walking time then was, in all honesty, little short of chaos. Whenever we went out, the dogs would all run around me, pulling on the leads. The situation summed up the fatal flaw in traditional training in many ways. I think I had instilled a lot of good habits into them through obedience training, but if I am honest with myself they were either robotic when we went out or doing their own thing—it was either everything or nothing. I didn't want that, and felt there had to be a way of achieving a kind of cooperation, a situation where I could get them to comply when I wanted and they could enjoy the freedom to run where they liked when they were able to do so. I knew the best form of control was self-control. But how to instill it?

Instead of putting them on a lead and letting them bounce around like maniacs, I thought I'd calm it right down again. As I was doing more and more now, I stopped and thought about the wolf pack analogy. I saw how the Alpha pair allowed the subordinates to run around for a while but that eventually all calmed down and they were able to lead the hunt in an orderly fashion. So the first time I gathered the dogs together for a walk, I did not try to stop them from getting excited—quite the opposite. Again, thinking about the principles of the wolf pack, I realized dogs have got to get wound up because, to them, this is the prelude to a hunt and they have to get their adrenaline pumping. What I was trying to do was not fight their instinct but go with it.

The difference this time, however, was that after putting the leads on the dogs, I did nothing. I just stood there, impassively waiting, calmly and silently, before heading out of the door. Again, the calming leadership I was showing bore fruit, and the dogs calmed right down. I then found that, on the walk, I had to keep showing them my leadership credentials. Previously, like so many other dog owners, I would be taken for a drag down the road by the dogs, an experience I never particularly enjoyed. However, I found that if, whenever the obligatory pulling started, I waited, the results were remarkable. The dogs quickly realized they were getting nowhere fast, and one by one their leads all slackened as they gave up trying and turned around to look at me. This was the first time they had done so, and it gave me the encouragement I needed to continue in this vein. It had been a battle of wills, and I had won them over.

I then started to wonder if the same approach would work when they were off the lead. In the past, my dogs would scatter to the four winds and then display "selective hearing"; they would come back to me perfectly well on some occasions, but if distracted by a rabbit or another dog, my futile attempts to call them back would echo across fields. On other occasions, I have seen dogs go back eventually, only to be smacked by their frustrated owner. I always thought that this was a confusing signal for the dog—surely it would make a dog wary of returning if it knew it was going to get clobbered? And if anybody has tried to catch their dog to get it under control, they know they can sometimes be led a merry dance by the dog, who waits for the owner to get close, then runs off again.

Once more, looking to the wolf pack gave me my answer to the selective hearing problem. Knowing that the Alpha wolf leads the pack on the hunt, I looked at the situation from the dog's point of view. If that dog believed it was Alpha, then it would think it was leading the hunt. Therefore, the owner's job, as subordinate, would not be to call the dog back, but to follow as a pack member. Encouraged by the positive response I had got working on the leads, I decided to show my dogs that I led the hunt off the lead as well.

I was not keen to test out this theory in an open field, but luckily I had enough room in my garden to make a start. Calling the dogs to heel and rewarding them for doing so immediately took away the confusion that arises when owners punish their dogs for coming to them late. Again, the dogs were quick to learn, all except Kim, the beagle. On one occasion, she was still not responding, preferring to nose around the garden. Frustrated, I turned away and headed for the back door, determined to leave her out there. As I reached the door and looked back, I saw Kim running flat out to get indoors. Inspiration struck. From then on, if Kim did not come when I asked, I turned around and walked back to the house, whereupon she would follow me. Dogs are, by nature, pack animals, and given the choice of going alone or returning to the pack, they choose the pack every time.

It was a huge leap forward. It was as if I held the dogs on invisible leads attached to them. The difference was astounding: within a week or so again, they were still enjoying their freedom, but now they were doing so in a way that meant they never strayed very far from me. And when I wanted the pack to reform to return home, they accepted the minimal instruction I gave to them instantly. I was, I must admit, over the moon.

* * *

I wouldn't want to create the impression that all this came easily, that everything fell into place instantly. It didn't, I can assure you. As I tried to develop my ideas, some things simply didn't work. In particular I found that any attempt to combine my new practices with the old, obedience training stuff did more harm than good. But as I thought about incorporating things like discs, clickers, and head braces, I realized "this is simply confusing." And if I was mixed up, what on earth would the dogs' response be?

I realize now that I was being human; I was overcomplicating things. I kept thinking, "There has to be more to it than this. It can't be this simple," and kept looking for other things. Slowly, however, it was dawning on me that in some ways it really was this simple. If I just concentrated on the dogs' way rather than the human way, I was going to be far more successful. It was obvious really. When do you ever *see* one dog using collars or leads or clickers on another dog? From then on, I determined that I was going to try to do this without resorting to any artificial man-made means.

By now I had been applying the principles with great success for two or three months; but a part of me was still convinced I was not getting the full picture. My own dogs were providing me with information on a daily basis, and as they did so, I was able to make little refinements to the techniques I was developing—it really was a question of trial and error at times. But the next big breakthrough did not come via the dogs I had then. Once more, it was my memories of The Duchess, Donna, that provided the inspiration.

I have always believed in treating my dogs to a supply of fresh marrow bones once a week. When Donna was around, the moment I put the bones down on the ground marked the moment the same little ritual would begin. In her usual imperious way Donna would walk silently in and the others would immediately stand back. Donna would then slowly sniff out the bones she wanted, then walk away with them. Only then would the others take what they wanted. It was, I realized, the same principle of leadership with which I was now so familiar. The one who appeared to do nothing got everything it wanted. And it made me think about using feeding time as a way of re-establishing the leadership structure. This was not a new idea entirely. The importance of eating in front of a dog was something I had read while studying the behaviorists. They recognized it as a simple way of showing them you are the leader. Again, this made sense to me having watched other animals, lions and—again—wolves in particular. It is always the Alpha that eats first in group feeders.

But while I agreed with the behaviorists' idea, I disagreed with the method that flowed from this. The behaviorists' approach was to impose a pecking order during the evening meal. Under this system, the human finished their meal in full view of the dog before allowing it to eat its meal afterward. It was a procedure that undoubtedly produced results, but there was a lot I was not happy about. Apart from anything else, people feed their dogs at different times of the day and night. Dogs in sanctuaries, for instance, are fed in the morning. I also thought the approach was too protracted. Again, I thought about dogs in the wild, and couldn't see how the pack would wait until the evening. A dog is an opportunist eater rather than only a gorge eater. It will catch a hare, a bird, any prey that will keep it going—it will not lounge around all day; getting food is the priority of the day.

On top of all this, it seemed an unkind thing to do. I put myself in the dog's place. I thought: "If you've gone all day without food and then the human sits down to eat before you finally get yours, you are going to be ravenous." This might put the dogs in their place but it is not very nice. I knew feeding time had huge potential as a means of reinforcing the leadership signals, but I wasn't going to eat a full breakfast or an evening meal in front of them, so I had to think of something else to get that information across. I had to come up with a new method.

I was beginning to realize that quick, instinctive information was the most useful, probably because a dog has no concept of the future at all. I had seen that sometimes the slightest gesture is capable of conveying a huge amount of information. The thought came to me one day. That evening, before I mixed their food, I put a cracker on a plate. Then I got out their bowls and mixed it up on a raised surface. What I then did was take the cracker out and eat it, making it look as if the food was coming out of their bowls. Again, I was thinking of it in terms of the pack mentality. What do they see? They see you eating out of their bowl. What does that make you? The leader.

I was not tackling bad behavior in this case. There were no particular problems at feeding time; quite the opposite in fact, it was a time when I knew I could get their undivided attention and their best behavior too. I fed them in their individual bowls, each of them dotted around the kitchen and the hallway. They knew their spots and—apart from their habit of exploring each other's empty bowls—behaved very well. In this case, my motivation was simply to underline the message I was getting across in the other areas.

They quickly sensed something was different. I can remember them looking at me rather strangely, trying to work out what I was up to. There was a little drama at first. There would be a little jumping and whining, but soon they were used to the ritual and would wait patiently while I ate my cracker. They seemed to accept that I had to be satisfied before they too could eat. Then when I placed their bowls down they ate contentedly. The changes were not dramatic, but on this occasion I had not expected them to be. It was simply another confirmation that I was their leader, another trick up my sleeve. And what pleased me most once more was that success had come by thinking of the nature of the dog.

By now I must admit I was feeling quite pleased with myself. Life always has a habit of cutting you down to size, however, and I was soon reeling from a terrible setback. I had already lost Sandy in the summer of 1992, but then in February 1994, I lost my beloved Khan. It was, I have to confess, a real blow to me. More than any other dog, Khan had been with me through good times and bad. I only had Sasha and the beagle, Kim, left. I missed the dogs I had lost terribly. It took the arrival of another dog to solidify all the ideas I had been working on.

5

The First Test

A few weeks after Khan's death, I popped into a local animal sanctuary. I had gone there to see the boss, a close friend, but my visit had nothing to do with dogs. It was about going to the theater, if memory serves me well. My friend was busy so, while I waited, I decided to take a walk around the sanctuary. As I did so, I came across one of the most pathetic sights I've seen in my life. Inside one of the blocks there was this thin little Jack Russell. I was aware of their reputation for being snappy and aggressive ankle biters, and had never particularly warmed to the breed. But it was impossible not to be drawn to this poor creature. He was trembling, and not just because it was winter and he was cold; I could see the fear in his eyes.

I soon learned his heartbreaking background. He had been discovered abandoned, tied to a concrete block by a piece of string. He had not eaten for days and was emaciated. If he had not been taken in by the sanctuary, he would have been dead by now. He was clearly a badly damaged dog. As I spoke to the kennel girl who was looking after him, she told me he kept running off. They were also worried that he might bite. Finding a new dog had been the last thing on my mind as I had driven over there. Nevertheless, I drove back with a new addition to the family shivering in the back seat. I had decided to take him in.

I soon named him Barmie, for no other reason than the fact that he was, well, a little bit barmy, mad. When I got him back home, he sat under my kitchen table. Every time I walked past him he growled. All I could feel was sympathy. It wasn't aggression I was seeing; it was nothing but sheer terror. I knew I'd be petrified if someone had treated me the way he had been.

I hadn't taken Barmie in as an experiment, but I was soon thinking that he was going to provide me with a great opportunity. I had so far been working with dogs that were comparatively well adjusted—animals that were used to always being treated kindly. Here I had one who had known nothing but bad treatment. Over the coming weeks, Barmie would provide me with the chance to test the knowledge I had been gaining so fast with my own dogs, to put all the pieces together. In return I hoped I would have the opportunity to help this troubled little dog get over his past.

By now a golden rule had begun to emerge: whatever it was that the traditional methods of training recommended, I needed to do the opposite. So I resisted the temptation to throw myself at Barmie, to shower him with love and affection. He was such a vulnerable creature that it was almost impossible at times. There were days when I just wanted to cuddle him and tell him he was all right. But instead I decided not to invade his space and just to leave him alone. So he just sat there under the kitchen table glaring. And I just carried on around the house as normal.

In everything I had read and seen, it was agreed that it takes forty-eight hours for a dog to suss out its environment. Then it takes about two weeks for it to sort out its place in its new home. It's like anybody starting a new job; it takes you about a day or two to sort out your desk, then another two weeks to find your place in the company. So for the first two weeks I continued in this vein, effectively leaving him to his own devices. Whenever I did speak to him, I did so as kindly as possible. Every now and again I would look across the room at him and just say: "Hello, love." I would see his little tail wagging, almost against his will as if he couldn't help it. It was as if he wanted to know what was required of him, but again I let him be.

The first thing I tried out with him was the "gesture-eating" technique. At this stage I was still experimenting with that theory. It was an ideal opportunity to really try it out because I had him on four very small feeds a day in an attempt to build him up. The poor thing had been starved, and weighed about two-thirds the weight he should have been. He responded immediately. He would sit there watching me with his ears pinned back. Then his little tail would begin to wag as if to say: "Yes, I've got that." I would then put his food down and walk away. He would watch me go, then tuck in.

He began to put on weight and slowly but surely he began relaxing. The growling stopped, he began slinking out into the garden when I was there hanging out the washing. Sometimes when I was sitting around, he would come up to me very, very tentatively. When he came, I wouldn't touch him, just let him get to know me. He was still very sensitive. When I got a lead out, he nearly died a death—if you become attached to a lead, you lose the ability to take flight. But I was not going to push him in any way, so I left it. My general principle remained that I was going to leave him alone, to give him time to make up his own mind.

The breakthrough came after about a month or so when I was out in the garden playing ball with Sasha. It was springtime now, and I remember that Sasha was retrieving and bringing a ball to me. Suddenly Barmie appeared in the garden with a rubber ring, or quoit, in his mouth. He had decided to join us. He was seeing that Sasha was getting attention because we were playing this game, and came over with the ring. I asked him to leave it and he did. I picked it up and threw it and he chased after it, grabbed it, then shot back into the house to hide under the bed.

I knew this was an opportunity to establish some sort of pattern so I decided I wasn't going to chase after him. I wanted him to play the game by our rules so I carried on playing with Sasha. Sure enough, a few minutes later he reappeared. He came up with the ring again, I threw it again, and he went to retrieve it again. This time, however, he came back and brought it back to me. I rewarded him with a "good boy" and repeated the exercise. He did the same thing again.

Every dog, like every human, learns at its own pace. In this case, we were talking about a remedial animal, a damaged dog, so I knew it was going to be a slow process. Finally, however, the breakthrough had come. I now knew that he was a more confident little dog. He had learned that nobody would hurt him. He felt he was safe and I could move forward with him.

I had showed him that I would play with him but only on my terms. So then I started calling him to me. One thing I bore in mind was that dogs are, like humans, selfish creatures by nature. This can be a means of survival or simply for fun, but dogs are driven by the question "Why should I do this?" My thinking was based on the idea of stimulus and reward, which I learned from behaviorism and B. F. Skinner, but by now I was adding to this the principles of the wolf pack and the primacy of the leader. I knew that the leader was not just the authoritative member of the pack but also the provider, so I too had to be both. Therefore when I called Barmie to me, I had a little piece of food in my hand. This began to work really well, so well that I then moved on to stroking him. Given how touchy he had been when he first arrived, this was a far more significant moment than usual. I could have cried when he responded to the affection. How long had it been since he had been shown such warmth, I wondered?

It was as I began to stroke him that I realized how far I had come. I noticed with Barmie that he would duck before I stroked him at the back of his neck. I had spent time with other dogs at the sanctuary and dogs there did the same thing, they ducked. My dogs didn't, and I wondered why did this dog act this way? When I did more research, I discovered that this is the most vulnerable area in most species, including humans. How many humans do you allow to touch your head and neck? Only those you trust. When dogs fight, the violence will begin when one arches over the neck of the other one. It was at this moment that I remembered something that Monty Roberts had said. He explained that if the animal believes in you, you can touch its most vulnerable area. It is, in some ways, the ultimate expression of your leadership. You are telling your subordinate that you know how to destroy it. The fact that you do not only underlines your authority even more. It made me realize how much trust I had now gained, how effective I had become in persuading my dogs that I was a leader to whom they could entrust their lives. It was a poignant moment.

My other dogs, Sasha and Donna in particular, had taught me a lot. But in terms of putting flesh on the bones of the ideas I was working on, Barmie was my best teacher by far. He taught me that I could not move on until he felt secure and comfortable and trusted me. There was no pain, no fear in him; he was now learning because he wanted to and he believed in me. He had also helped me see that all the elements of my method must happen simultaneously. It is an absolute circle of events and the dogs must have that information supplied to them consistently.

The events of the last few months had been exciting and incredibly rewarding. I can't begin to tell you the calm that had come over the dogs; it was awesome. And the more I took charge of these situations, the more control I had, the more I had them willing to do what I wanted. What made it even more rewarding was the fact that there was none of the enforcement of so-called obedience work. I had finally proved what I had sensed for so long: that it was possible for dogs to follow me because they wanted to rather than because they had to.

Predictably, the reaction of humans was less serene. By now I was talking openly about what I felt I had achieved—to a mixed reception. Some people smiled sweetly, shook their heads slowly, and gave me a look that told me they thought I had finally gone off my trolley. Some people were more outspoken. Some said, "Oh, you are cruel." Others dismissed me with "Oh, you and your daft ideas." I'm not going to pretend I'm made of steel. I was hurt a lot. A couple of times I thought to myself: "Why am I making all this trouble for myself? Why am I bothering?" But again I thought of Monty Roberts, whose father had beaten him for his ideas as a boy and who for almost forty years had put up with the scorn and ridicule of the horse world. I thought if Monty could stick to it, then so could I. Perhaps unsurprisingly, of those who understood what I was up to, it was Wendy, who had after all introduced me to Monty Roberts, who was one of the most supportive. She was adopting my principles and trying them out on her dogs with encouraging results. She told me to persist, keep at it.

Slowly but surely the word spread and people began to ask me how it could work for their problem dogs. I began to visit people, applying the techniques I had learned with my own dogs to their problem pets. Seeing was believing. In home after home that I visited, the dogs' behavior would change immediately. I saw that the dogs were free and happy to change, that they wanted to do it. It was powerful stuff, and I felt very humble, very privileged.

Six years later, I have worked with hundreds of dogs. The communication technique I have evolved has helped improve the behavior of all of them. I have now reached the point where, if an owner does as I say, their dog will have to do what that owner wants. The principles I laid down during those exciting early days now form the basis for my work.

6

Amichien Bonding: Establishing Leadership of the Pack

No one could have a higher regard for the intelligence of the dog than me. There are still times when I seriously wonder whether they are wiser creatures than some of the humans with whom I come into contact! Even I have had to accept that one thing is beyond them, however. Dogs are never going to learn our language. The bad news is that to communicate successfully with our dogs, it is up to us to learn their language. It is a task that requires an open mind and a respect for the dog. No one who regards a dog as their inferior will achieve anything. It must be respected at all times for what it is.

The good news, however, is that whereas humans speak in a bewildering range of tongues and dialects, dogs share one universal language. It is a silent and extremely powerful language, yet at its heart are a simple set of principles that—with a few subtle variations—influence the way all dogs behave. To understand the principles of this language, we first have to understand the society within which all our dogs believe they are living. And the model for this community is the wolf pack.

The modern dog's appearance and lifestyle is, of course, far removed from that of its ancient ancestor. Centuries of evolution have not removed its basic instincts, however. The dog may have been taken out of the wolf pack, but the instincts of the wolf pack have not been taken out of the dog. Two immensely powerful forces guide the life of a wolf pack. The first is its instinct for survival, the second its instinct for reproduction. The means it has evolved to guarantee these ends is a hierarchical system as strict and successful as any in the animal world. Every wolf pack is made up of leaders and subordinates. And at the head of every pack's pecking order are the ultimate rulers: the Alpha pair.

As the strongest, healthiest, most intelligent, and most experienced members of the pack, it is the Alpha pair's job to ensure its survival. As a result, they dominate and dictate everything that the pack does. Their status is maintained by consistent displays of authority. Underlining this, the Alpha pair are the only members of the pack who breed. As humans we have, of course, developed along different, what we would like to believe are more democratic, lines. Yet sometimes I wonder whether it is we rather than the dogs who took a wrong turn. How much trust can we really place in our leaders? How many of us have even met them? Within the wolf pack no such uncertainty exists. The Alpha pair control and direct life within the pack and the remainder of the pack accept that rule unfailingly. Each subordinate member is content to know its place and its function within this pecking order. Each lives happy in the knowledge that it has a vital role to play in the overall well-being of the pack.

The hierarchy of the pack is constantly reinforced through the use of highly ritualized behavior. The ever-changing nature of pack life—in which Alphas and their subordinates are frequently killed or replaced through age—makes this essential. As far as the wolf's modern-day descendants are concerned, however, four main rituals hold the key to the pack instinct which lives on within them. They are central to all that will follow.

It is no surprise to discover the Alpha pair are at their most dominant during hunting and feeding times. Food, after all, represents the pack's most fundamental need, its very survival depends on it. As the strongest, most experienced, and intelligent members of the pack, the Alpha pair take the lead during the search for new hunting grounds. When prey is spotted, they lead the chase and direct the kill. The Alpha pair's status as the pack's key decision makers is never more in evidence than during this process. The wolf's prey can range from mice to buffalo, from elk to moose. A pack may spend hours stalking, cornering, and slaying its target, covering as far as fifty miles at a time. The organization of this operation requires a combination of determination, tactics, and management skills. It is the Alpha pair's job to provide this leadership. It is the job of the subordinates to follow and provide support.

When the kill has been made, the Alpha pair get absolute precedence when it comes to eating the carrion. The pack's survival depends on their remaining in peak physical condition after all. Only when they are satisfied and signal their feed is over will the rest of the pack be permitted to eat—and then according to the strict pecking order, with the senior subordinates feasting first and the juniors last. Back at the camp, the pups and babysitters will be fed by the hunters' regurgitation of their food. The order is absolute and unbreakable. A wolf will act aggressively toward any animal that attempts to eat before it. Even the fact that the pack contains its blood relatives will not stop the Alpha attacking any animal that breaks with protocol and dares to jump the queue.

The Alpha pair repay the respect the pack bestows upon them with total responsibility for its welfare. Whenever danger threatens, it is, once more, the role of the Alpha pair to protect the pack. This is the third situation in which the natural order of the pack is underlined. The Alpha pair perform their leadership role unblinkingly, and from the front. They will react to danger in one of three ways, selecting one of the "three Fs"—flight, freeze, or fight—and will run away, ignore the threat, or defend themselves. Whichever response the Alpha pair select, the pack will again back up their leaders to the hilt.

The fourth key ritual is performed whenever a pack is reunited after being apart. As the pack reassembles, the Alpha pair remove any confusion by reasserting their dominance via clear signals to the rest of the pack. The pair have their own personal space, a comfort zone if you like, within which they operate. No other wolf is allowed to encroach into this space unless invited to do so. By rejecting or accepting the attention of other members who wish to enter their space, the Alpha pair re-establish their primacy in the pack—without ever resorting to cruelty or violence.

We may consider them to be pets but our dogs still believe they are functioning members of a community that operates according to principles directly descended from the wolf pack. Whether its "pack" consists of itself and its owner, or a large family of humans and other animals, the dog believes it is part of a social grouping and a pecking order that must be adhered to at all times. What is more, all of the problems we encounter with our dogs are rooted in their belief that they rather than us, their owners, are the leaders of their particular packs.

In our modern society, we keep dogs as eternal puppies, feeding them and caring for them, so they never have to fend for themselves. This is why dogs should never be given the responsibility of being Alpha of a pack, as they will simply be unable to cope with the decisions they face. The responsibility puts immense pressure on them and leads to the behavioral problems I so often witness.

In the course of the last few years, the many dogs I have worked with have suffered symptoms ranging from biting to barking to bicycle chasing. Yet in each and every case, the root of the problem lay in the dog's misplaced belief about its place within the pack. So in each and every case, I have started the same way, by going through the process of Amichien Bonding. I have never deviated once; it is absolutely fundamental.

The bonding takes the form of four separate elements. Each correlates to the specific times I have identified when the pack's hierarchy is established and underlined. On each occasion, the dog is confronted with a question which we must answer on its behalf.

- When the pack reunites after a separation, who is the boss now?
- When the pack eats food, what order do they eat in?
- When the pack is under attack or there is a fear of danger, who is going to protect us?
- When the pack goes on a hunt, who is going to lead us?

This is a holistic method of working; all four elements must take place in conjunction with each other, and they must be repeated constantly, day in, day out. The dog must, in effect, be blitzed with signals. It needs to learn that it is not its responsibility to look after its owner, that it is not its job to care for the house, that all it has to do is sit back and lead a comfortable and enjoyable life. It is a mantra that must be repeated over and over again. Only then will a dog get the message that it is no longer in charge; only then will it be able to exercise the most powerful form of control, self-control. After this has been achieved, the task of tackling the specific problems of the individual dog becomes infinitely easier.

1. Reuniting: The Five-Minute Rule

The first requirement of Amichien Bonding is to establish leadership during day-to-day life at home. To do this involves tackling those moments when the dog and its owner are reunited after a separation. Most people imagine these reunions happen on a handful of occasions each day, when they go out to work or to the shops for instance. In fact the act of separation occurs on countless occasions every day.

Throughout all that is to follow, the dog should be seen, not as a lovable domestic pet, but as the deeply protective, fiercely loyal leader of a wolf pack. So, regardless of whether its owner leaves the house or simply leaves the room to go to the garden or the bathroom, the dog sees it as an instance of a charge or child leaving its protective custody. While the human probably knows how long they will be absent, the dog does not. As far as it is concerned, its charge may never return and it may never see them again. So whether they are away for eight hours or eight seconds, the moment the charge reappears the dog will go through a ritual aimed at re-establishing its leadership. To redress this, the owner must begin displaying the behavior of a leader. And the first step to establishing that leadership is learning to ignore the dog.

All dogs go through their different rituals when they are reunited with their owners. They may start leaping around or barking, licking or bringing in toys. Whatever they do, it is crucial that the owner turns a blind eye, that he or she pretends it is not happening. Failure to do this means the dog has been acknowledged, or paid homage to—that its behavior has succeeded in getting attention and the dog has achieved what it wants. Its primacy has been confirmed. Even by turning around and saying "stop it," an owner is allowing a dog to achieve its aim. The key to this then is that the dog must not be engaged with in any way. By this I mean no eye contact, no conversation, no touching unless it is to gently push the dog away. The owner must do nothing.

No matter how agitated or aggressive the dog, it will at some point decide to bring this ritual to an end and walk away. In most cases, the dog will probably take a brief time out to evaluate what has happened. It may very well return and go through the same repertoire again. If it does, ignore it. What is happening is that the dog is sensing a fundamental change in its environment. Each time it returns, it does so trying to spot a chink in the aspiring new leader's armor. I have seen dogs go through the same ritual a dozen times before giving up. Each time the performance becomes more and more muted. By the end, their bark may be barely audible. The key thing to remember is that nothing can happen until this repertoire is over. Any attempt to get the dog to cooperate with you before then will be futile.

The dog will signal that its resistance is over by relaxing or walking off somewhere and lying down. It is the first indication owners get that the dog is seeing them and their relationship in a new light. The dog's deferral reflects a new respect for the owner's space. The process is far from over, but an important breakthrough has been made.

The important thing now is that nothing happens for at least five minutes. The dog can be given more time if preferred, but under no circumstances should anything else be attempted before those five minutes have elapsed. I call it "time out." During this time, the owner should just carry on with their normal routine. Some get impatient, so I tell them if they can think of nothing else, they should pop into the kitchen and make a cup of tea or coffee, as that usually eats up the time. The object of this break is to allow the silent process of deposing the dog to begin. What the owner is inviting the dog to do during this time is to dwell on what has just happened. It is being given space to realize two things have occurred. Firstly, its ritual has failed to achieve any sort of response and, secondly, something has changed within its relationship with its fellow pack member. There has been a subtle shift in the pecking order.

Some dogs are quicker on the uptake than others. In some instances it may take less time, in other cases it may take longer. In my experience, however, five minutes is generally long enough for this assimilation to take place. If during that time a dog comes to its owner uninvited, it must be ignored; even if it comes to sit on its owner's lap, it must be ejected without a word. The dog must not be allowed to demand anything anymore.

It can, of course, be a challenge, particularly with big, physical dogs. But an owner must be steadfast. If an owner is standing and the dog comes at their body, they must block it with the body and turn away from the dog. If a dog jumps up, putting its front legs onto the owner's lap, the owner must silently put a hand on the dog's chest and push them down gently. The owner must not shove or say anything; I cannot emphasize this point enough. Even saying "go away" ensures the dog has got its way and has been acknowledged. Once the five minutes have elapsed, the job of engaging with the dog can begin. And it is by engaging with it in a specific way that an owner will be able to underline the message that a new leadership has been established.

I often hear people complain that it is cruel to ignore a dog in the manner I advocate. My response is always the same: the fact is that by establishing my relationship with the dog on the correct footing, I can enjoy its company even more. By allowing myself the time to get on with the other jobs I have to do at home undisturbed, I am able to make the time I spend with my dogs real quality time. All owners can begin making that quality time for themselves from the very beginning. I am not saying for one moment that owners should ignore their dogs from now on; they can still fuss and pet over their companions as much as they want, but on their terms. The dogs will be happier in this type of relationship, as there is no confusion as to who looks after whom.

The Come

Once the five minutes have elapsed, an owner can begin interacting with their dog according to the new rules. And the first task I ask them to practice is getting the dog to come to them when they want. The principles guiding them here are request and reward. I use the word "request" rather than "command" advisedly because what we are talking about here is a two-way street. Always remember, we are trying to create a situation where the dog is doing things of its own free will. We want the dog to elect the owner as leader of its own volition.

The key points I ask people to remember as they move on is that they should always make eye contact and should always call the dog by its name. Most importantly of all, they should always remember to reward its good behavior when it does come as requested. The choice of reward is entirely the individual's choice. Small pieces of cheese or chopped liver or meat strips make very effective tidbits, I find, but here it is up to each owner—use whatever your dog likes. A woman once asked me whether she could give her dog a whole tin of dog food. Given the amount of rewarding involved in the early stages of the process, that would produce a rather overweight dog.

The important thing is that the second the dog comes, the reward is offered in the dog's mouth and that the dog is told "good boy" or "good girl." I also suggest owners gently stroke the dog's head and neck. From the very beginning, they are establishing an important principle: the dog has done what it has been asked to do and the minute it has done so it has benefited. By rewarding the dog with food, repeating praise, and stroking the dog in a hugely important area of its body, the owner is sending out a powerful message that will be replicated time and time again from now on. If the dog comes to the leader when it is asked, that leader will reward it.

This is a crucial stage in the early establishment of an owner's leadership and should be practiced until the response is exactly what he or she requires. It is quite possible, for instance, that the dog may respond to the attention and the stroking in particular by becoming agitated once more. If the dog starts to slip back into its old ways like this, the owner must at this point stop immediately and leave the process for at least an hour before starting again. The dog must understand that there are consequences to its actions and just as good behavior is rewarded with food or affection, undesirable behavior produces a less enjoyable consequence: it loses what it craves most, its leader's attention. If this does happen, I ask owners to simply repeat the process from the beginning and keep on repeating it calmly and consistently until the dog understands what they want. It is vital that owners don't rush, and most of all, don't become angry. I ask them to keep their pulse rate low at all times; I tell them to remember Kipling and keep their heads.

An additional tool in this stage is the creation of "no-go areas" within the home. Early on, a dog can be taught that certain areas of the house belong to the leader. Again, the dog will recognize the principles at work from its instinctive connection to the wolf. Within the pack, the Alpha's space is respected at all times. Subordinates only enter this space at their leader's invitation.

Hopefully, a dog should respond immediately to the new system. If it does, the owner simply needs to spend the next few days going through the same process again, beginning and ending it the same way. As they progress, they should notice the dog beginning to respond to the call of its name without rushing. This is a good indicator that they are approaching their goal. I liken the behavior of a dog who has grasped my method to that of a well-behaved child responding to the authority of a schoolteacher. Asked its name in class, a child will acknowledge the teacher then wait to be given its task. I want the dog to behave in precisely the same way. I want it to stand or sit there, acknowledge its owner with eye contact, and then await their request, whatever it may be.

Dogs have lots of wonderful qualities but they are not—to my knowledge anyway—mind readers. They do not know what we want of them. By laying down this groundwork, by establishing leadership in this way, owners are paving the way for a new relationship. From now on, a dog will no longer have to guess what its owner wants. It is ready to listen to and cooperate with its owner's requests. It is also ready to relax and enjoy life.

2. Danger Signals

One of the messages I emphasize when I am working with owners is that all the four elements of Amichien Bonding must act in conjunction with each other. As they begin the first part of the bonding process, they should also begin dealing with a second key area, what I categorize as moments of perceived danger. This most commonly manifests itself at home when visitors arrive. We have all witnessed dogs going berserk at the sound of a doorbell or knocker. There is not a postman or milkman alive who has not been at the receiving end of this sort of unwanted attention. Again, the key to understanding this behavior lies in the pack. If a dog believes it is the leader of its pack, it believes it is its role to defend the pack's den. So in instances like this, the dog is responding to an unidentified threat. Someone or something is about to enter its community and it is anxious to know precisely who or what it is. It believes it is then its responsibility to deal with the intruder.

There are two elements to the process I ask owners to go through here. The first involves the owner, the other the visitor. When the dog begins barking or jumping up at the sound of someone at the door, the job of the owner is to thank the dog. The point here is that the owner, as the leader, is acknowledging the vital part the dog is playing in the pack. The dog has realized there is potential danger and has alerted the decision maker. It is like a child that has told its parents there is someone at the door and has been thanked for doing so. Relieved of responsibility, the dog can get on with leaving the decision maker to decide whether this visitor will be allowed through the door.

All dogs are clearly different. Some will have developed worse habits than others so there will inevitably be different reactions—from both the dogs and the humans. Experience has taught me that there are four ways of approaching this situation. Firstly, owners can permit the dog to come to the door with them. If this is the case, however, the guest must be asked to completely disregard the dog in the same way that the owner has been doing after separations. It must be explained to them that whatever their instinct, they must not fuss over the dog.

This is, I know, very difficult, particularly for those who love animals and in the case of dogs that are right up there in people's faces demanding attention. So the first alternative to this is to offer the owner the option of putting the dog on a lead. This will allow him or her greater control if the situation becomes difficult.

If the dog's behavior is truly unacceptable, the next alternative should be applied and the dog should be asked to go into another room. It is vital that this is not seen as an act of exclusion or punishment, however. The dog must not be physically shoved or lifted out of the way. It should not be thrown out of the house, into the garden for instance. Throughout the process I want the dog to be making positive associations with its behavior in certain situations. So this should be done according to the reward principles already established. The dog should be praised for recognizing the danger, then removed from the decision-making process and given a favorite tidbit for cooperating. The door can then be shut so that it is out of the way temporarily.

By dealing with the situation in this way, the owner will create the time and space to tell the guest what is going on. The visitor can then be briefed to behave in the same way as is now the norm. Once this is understood, the dog can be safely let back into the living area. I always ask owners to be sure the dog does this without anyone speaking to them as they enter. If this happens, the dog should recognize the situation is normal and begin behaving as it has been.

The fourth and final option for dealing with this area applies if a guest either doesn't believe in what an owner is doing or simply cannot understand it. Children, of course, are the most obvious example of the latter and I will deal with them in detail in due course. In this case, it may be best to leave the dog in a separate room. This may be the best course of action too if you have friends and family who simply will not go along with the process. For most people, it is not worth falling out with friends and family over.

Basic Controls

In many ways, learning Amichien Bonding is comparable to learning to drive a car. In time the fundamental routines will become second nature. It will only be in challenging situations that owners will even need to think about the practices they are applying. For the most part the knowledge will be stored away in the subconscious, a useful new skill that will add enormously to the enjoyment of the dog-lover's life.

No one, however, is allowed to drive a car without being shown where to find and how to operate basic controls such as the brakes, clutch, and accelerator. The next stage an owner will, however, move on to is walking the dog. Before he or she is able to move out into the wider world, the owner must learn the basic skills required to exercise control in that world. As in all dog training regimes, these controls are the ability to get a dog to come to its owner, to walk at its owner's side, to sit, and to stay.

There is, as the old saying goes, no place like home. And when it comes to laying down the foundation stones of my method, this is certainly the case. I passionately believe there is no place like the dog's own environment to begin building the relationship established by Amichien Bonding. So I ask that owners allow a fortnight at least to bring all the elements of my method together.

Of course, the process of getting the dog to come at its owner's request has already begun during the bonding work that follows the five-minute rule. At this early stage, the dog has begun to realize that certain behavior is rewarded by food while other behavior is not. It quickly chooses the behavior that benefits it the most. This principle will remain central to every element of training at every stage.

As they move on, the first thing I recommend owners do is teach the dog to sit. It is, for most ordinary dog owners, the most important means of getting a dog to exercise its right to freeze. It is a useful—and at times vital—control to have available. In certain dangerous situations, it can save a dog's life.

Central to everything I do is the idea that dogs begin making choices of their own free will. At every turn I want them to make positive associations with certain behavior. I want them to recognize the situations where they know there is something in it for them, that instinctively they will be rewarded if they do the right thing. As I have already said, there is no more powerful tool in this respect than food. To teach a dog to sit, I ask the owner to first call their dog to them, then to bring a morsel of food up to the dog, almost touching its nose, then draw the morsel over the dog's head. As the dog instinctively arches its head back to follow the smell, so its body will tip back naturally as well. As this happens, the dog's bottom should touch the ground. As it does so, the tidbit should be popped in the dog's mouth, instantly accompanied by a verbal confirmation: the word "sit." The signal is clear; the dog's action is good and it is being rewarded.

If the dog moves backward when following the morsel, a hand can be placed behind it to prevent this. Hands must never be used to force the dog's bottom to the ground. If, for whatever reason, the dog moves away, I ask owners to simply remove the food from the dog's vicinity and start again. If this is repeated, the dog will quickly learn the realities of life: if it does the job right it will get its wages, if it does it wrong it won't. It will soon be sitting naturally. Dogs are highly intelligent creatures of course. If a dog begins sitting in front of its owner without being asked, it should not be rewarded in any way; the dog is attempting to regain control of the decision making.

From here, I recommend owners move on to heel work. By this I mean getting the dog to understand that the best position for it to be is at its owner's side at all times. I advise that this is again taught off the lead; the option of flight is then open to the dog if it gets frightened, so it will feel comfortable and secure. Again, food is the ideal means of communicating this message. I ask owners to encourage their dog to come to stand by their side using their pet's preferred tidbit. As in the other work, I ask owners to underline the message they are delivering to the dog by stroking. The key thing here again is that their stroking is confined to the key area of the head, the neck, and the shoulders. The signal is unequivocal: I am the leader, I know your weakness but I am here to protect you. The dog will have no option but to trust anyone who presents such formidable credentials.

In most cases, the ability to sit and remain at heel are enough. But I am a great believer in getting a dog to lie down on request as well. My reasoning here is simple. Calm is all-important in every element of my method and this is the most relaxed position a dog can take. I encourage the dog to do this again through reward and stimulus, but in this case by drawing the dog under a low piece of furniture, a table or a chair, and then getting it to lie down. Once more, I am manipulating the situation, getting the dog to do something for a good reason rather than using force. Again, it is an idea a dog picks up incredibly quickly.

A point worth making at this stage is that the dog does not need to be rewarded with a tidbit every time it completes a feat satisfactorily. Food is a powerful means of transmitting the initial message. As the process evolves successfully, however, I suggest owners reduce the frequency of food rewards gently. They might begin by dropping down to every other time the dog does the right thing then to once every half a dozen times until food is being given once every twenty times. The tool should never be removed from the process completely, however. It is important to keep the interest alive.

As in so many instances, the parallel with children is an appropriate one to draw here. I recall a moment with my granddaughter, Ceri, when she was being taught good manners by her parents. She had learned to say the magic word "please," but on one occasion failed to use it when asked if she wanted a drink. "I forgot, I'm only four," she said with an angelic smile. Dogs are no different. They too take their time to grasp things completely. Given time, affection, and encouragement, however, they will get there.

People often wonder whether my regime somehow removes the enjoyment from owning a dog. I always find this baffling; in fact the opposite applies. By removing the responsibility from the dog's life, an owner is ensuring it a happier more carefree existence. And by creating an environment in which an owner can interact with their dog at the times he or she chooses, the dog is being given quality time with its leader. That quality time can be used to build an even deeper and more rewarding relationship.

Two specific activities, toys and grooming, are particularly enjoyable in building the relationship owners are seeking here. Toys offer a perfect means to both bond with the dog and underline the leadership hierarchy at the same time. Equally, an owner can derive great pleasure from grooming his or her dog.

Again, the reward principle applies. If a dog allows itself to be brushed gently without protest, it can be praised and rewarded with food. These are all positive building blocks for the life ahead. I will look at both areas in a little more detail later.

3. Taking Charge of the Walk

The first disciplines—coming, sitting, and heel work—should in almost all cases take no more than a week. They provide the foundation for the next major area, going on a walk, which is equivalent, in a dog's eyes, to leading the pack on a hunt. People's walking habits will, of course, vary considerably. Some will only have the time to take their dogs for a short walk each morning and evening. Others will be free to go on long and frequent walks at any time of the day or night. My method is intended to fit into all lifestyles. Whatever the situation, the key to this element of the process is that owners take charge of the walk. By far the easiest way for the owner to know if the walk is going according to plan is to ask themselves if they are happy and in control. Once more, calmness and consistency are crucial.

The first task is to get the dog used to a lead. I personally prefer light rope leads. Chains, to me, seem like weapons, and if you bear in mind that a dog only pulls on a lead because it believes it must, being leader, any form of physical restraint will not change its mind. The dog's mind must be changed as to its role in the pack. I ask owners to call their dog to them. Then, using a food reward, place the lead over the head. This is without doubt one of the most intense moments of the method; it marks the first occasion when the dog has been denied the option to flee. It is also the first time the owner has placed an object around the immensely important head, neck, and shoulder area of the animal. If the dog shows any anxiety about this, make the association with the lead a positive one by using a food reward. Once it has accepted the lead, the dog's belief in the owner's leadership will deepen yet further.

It is, of course, hardly surprising that all dogs become excited at the prospect of heading off into the big wide world. To them, they are heading off on the hunt, the most elemental activity of all. The adrenaline rush they experience is welcome. It is the owner's job, however, to keep the dog's enthusiasm steady. It is an important test of leadership.

When the dog has accepted the lead, I ask the owner to get it to come to heel, again using a food reward if necessary. If the dog attempts to pull away, I instruct owners to stand still. The dog is being demonstrated the consequences of this action. The owner should then go back to the beginning and ask the dog to come to heel once more. Once the dog has come to the owner's side, it is time to move off. Again, any sign of pulling on the line must result in the lead being relaxed and the walk suspended. The crucial message that has got to be put across now is that the dog must remain close to and not in front of the owner, but at their heel. Any deviation results in a return to the den.

This principle is never more important than at the next crucial stage: as the owner goes out of the front door. To the dog, this is a portal into another world, an exit from the den and a home to a million potential threats. It is absolutely vital that the owner goes through the door first. This signifies that he or she is the leader and that he or she is performing the job of making sure the coast is clear. Again, this is an immensely powerful signal. If the dog somehow forces itself out first, then it is back to the beginning.

The themes established indoors must remain in place once the dog has moved outdoors. As the walk begins, for instance, the dog must never be allowed to walk ahead. Once more, that position is reserved for the leader. If the dog senses this position is acceptable, its belief that it is leading the hunt will be established. Instead, the dog should remain at the owner's side throughout.

Dogs can, of course, become extremely excitable at this stage. Pulling on the lead is one of the most common problems faced by dog owners everywhere. It is imperative that owners do not get into a pulling battle. Even the smallest dogs can pull really strongly. The game must not be indulged. The dog must play by its owner's rules, not by its own. If a dog pulls continuously, the lead must be relaxed, signaling the walk is not going to take place. To many, this may seem hard, but it will not last for long. When dogs learn that by pulling on the lead the walk doesn't happen, it doesn't take long for the penny to drop.

Of course, there are people who will argue that denying a dog its daily walk is cruel. To my mind, however, it is more important that it establishes total trust in you before stepping out into the wider world. Otherwise it is being cast out into an environment it does not understand and asked to perform a leadership role it is simply not equipped to fulfill. To my mind, this is far more cruel. And besides, whatever short-term sacrifices an owner makes in this time will seem minuscule in comparison to the huge benefits that should follow.

The Stay and the Recall

Walking a dog is, of course, one of the great pleasures of life. No owner can fail to enjoy the moment when their dog is released on its run, free to express its personality and natural athleticism. As they move on to this stage, however, I ask all owners to add two additional skills to their repertoire: the stay and the recall.

Dogs should always remain on the lead in built-up areas or near roads. It never ceases to surprise me how many people fail to realize the intrinsic danger of letting a dog run free in such hazardous situations. Once in open space, however, the dog can be readied for release. The first time this is attempted, I recommend owners go through a routine that once more underlines the principles established at home.

The first discipline is to teach the dog to stay. This is easily achieved by keeping the dog on the lead. The dog should be asked first to sit in the normal way. The owner should then turn and face the dog, take just one step backward while at the same time raising the palm of his or her hand and issue the request: "Stay." The dog should then be asked to come. This should be repeated, with the owner moving a little further away each time. If the dog moves off, however, it should be returned to the point where the process first began. Again, the dog must learn the consequence of its actions. The rules of this game must remain in the control of the leader.

With this extra control in place, the owner is ready to release the dog. When the lead is first removed, I recommend the dog be encouraged to stay at the owner's heel for a short while. As usual, a little food incentive can be used to make sure this happens. The dog should then be given a release word that it will recognize from then on: something like "Go play."

The key test now is to learn whether the dog is going to return. Again, this is done through response and reward. I suggest owners try to ask the dog to come back to the heel position as soon as it strays more than ten feet away from them on the first walk. The knowledge that it will return will help both the owner and the dog enjoy the run from then on.

Ultimately it is up to each owner to decide when and whether to allow the dog off the lead or not. It should not be attempted if there is the slightest worry that it may not come back. I advise anyone who is unsure of this to test the dog's response to a request to come, inside the home or in the garden. The response here will provide a good guide to how it will act in a more open environment. With dogs who prove difficult in this area, I recommend that an extension to the lead is added. This can be used as a way of helping the dog to understand what is wanted by gently drawing the dog to you with a request to come and a food reward.

4. Food Power

The controls applied in the wild by wolf packs are, of course, beyond us. Even if we wanted to, we are physically incapable of replicating the aggression and extraordinary body language with which the Alpha exerts its leadership. Yet by adding a little human ingenuity and subtlety, I believe one of the most potent tools available to the Alpha is available to us. Obtaining the power at feeding time is an immensely important element of Amichien Bonding.

For reasons that will be obvious, I call this element of the technique "gesture eating." It is an element I ask people to apply for the first two weeks or so only. If at all possible, I prefer every human member of the family to participate. By acting as a team, this will allow them to communicate an immense amount of information and establish each of them at an upper level of the household's hierarchy. Again, the overriding priority here is to be consistent, so it is essential this is repeated at all the dog's meal times during this period. Many people, for practical reasons I can understand, feed their dogs during the evenings only. For maximum impact, I prefer it if dogs are fed twice daily, once in the morning and again in the evening.

The technique is simple. Before preparing the dog's food, I ask owners to place a small snack—one per person in the home—on a plate on a raised working surface. Anything will do, a biscuit, a cracker, or a piece of toast. I then ask them to place the dog's bowl next to the plate. Making sure the dog is paying attention, they should then proceed to mix its meal. Once this is done, without speaking to or looking at the dog, each member of the family should reach for their snack and eat it. Only when everyone has finished eating their biscuit or cracker should the dog's bowl be placed on the floor. This should be done again with as little ceremony as possible, and only minor recognition of the dog. Then the owner should walk away and leave the dog to eat in peace.

The message here is clear and powerful. As in the wolf pack, the pecking order is clearly displayed at feeding time. It is the leader and its subordinates who eat first. It is only when they are satisfied that the next ranking member of the pack is able to eat its meal. To underline this message, any dog who walks away from its food during meal times must have its bowl removed immediately. Owners should not worry about it starving. When it comes to matters related to meal times, dogs will pick up the thread extremely quickly, take my word for it. The point here again is that the dog must learn that only acceptable behavior is rewarded. It is the leader who dictates the terms under which food is distributed and eaten. If it does not adhere to the leader's rules at meal times, it will miss its turn.

Dogs are pack animals; they like to live in groups. I often say to people that two dogs are half the work of one. They do play together, they do amuse each other, and when their owner is absent, they provide company for each other. Whatever the domestic set up, however, it is important to remember that the dog regards the other animals, including the humans, that share its space as fellow members of its pack. We all need to live by the rules and the dog is more willing to live by the rules than we are. The key to everything I do lies in establishing a set of rules that the dog will understand within the context of its pack. Once an owner has begun applying the four principles I have outlined here, it should take him or her around a fortnight to get their dog to digest those rules fully. Of course, no two pets are the same. The more damaged the dog, the longer it takes; the more severe the behavior, the longer it takes. There is no place for fear or pain in my method so my message is always the same: be patient, be gentle, and it will happen.

7

Separate Lives: Dealing with Separation Anxiety

From obsessive behavior to bedwetting or biting, I begin every case I see with the process of Amichien Bonding. Only when the dog's misplaced sense of status has been removed can it and its owners begin to lead a more relaxed and rewarding life. But of course no two sets of circumstances are the same and no two problems are precisely the same; in fact every dog I have worked with has demonstrated more than one problem, not just the problem the owners were concerned about. As a result, I have found myself adapting my method to deal with a wide range of dogs and an even wider range of problems. If one thing became apparent as soon as I began doing so, it was that my life was never going to be dull.

No case illustrates this better than one of the first I dealt with, that of Sally, a district nurse who lived in a lovely cottage in a village a few miles away from my home. Sally rang me one evening in an agitated state. "I've heard about the work you've been doing," she said. "I wonder whether you can sort out my Bruce." Bruce was a four-year-old mongrel, a handsome, fox-like chap. Sally loved him to pieces and he felt much the same way—the problem was he loved her just a little too much. And he simply could not bear to be parted from her!

When she was at home, Bruce would trail around after her

wherever she went. He would be constantly under her feet. Her real problems began whenever she left the house, however. The moment Sally went out of the door, all hell broke loose. Bruce would fly around the house, frantically grabbing hold of whatever pieces of Sally's clothing he could find. She would often return home to find assorted pieces from her wardrobe arranged in the form of a bed in which Bruce had been lying. Needless to say, her dry cleaning bill had become astronomical. Many of her favorite outfits were unwearable.

By far the most disturbing aspect of Bruce's behavior, however, was the way he had begun physically attacking the front door of the cottage. At first he had begun chewing at the wooden frame. His attacks had slowly stripped the wood away to expose the wall underneath. By the time Sally rang me, he had gnawed his way through the wallpaper and the plasterwork so that the bare brick had now been exposed. The doorway looked awful. Sally had become desperate to call in the local carpenter but knew there was little point fixing the frame until Bruce mended his ways.

In the years that have followed, I have seen these symptoms on countless other occasions. Bruce's behavior represented a classic example of one of the more common problems I have to deal with: separation anxiety. There is no question that being separated from its owner can be terribly upsetting for a dog. The sense of anguish the dog feels can be the cause of some terrible destructive behavior. I have seen dogs who eat furniture and curtains, clothes, and newspapers. I recall one dog who ate a cassette tape; it had to be operated on so that the spaghetti-like length of tape that had unspooled in its stomach could be removed. It hardly needs saying, but dogs can kill themselves in such situations.

Yet my experience has shown me that a dog's anxiety is not through pining like some abandoned child, but that it is the dog who sees itself as a parent, and it is distressed because its child is out of its sight. It did not take me long to realize that this was precisely what Sally's dog, Bruce, believed. It was soon clear too that the life the two of them led together only served to underline this situation. The first thing I noticed when I visited Sally was that Bruce came bounding up to me. This was clearly something Sally perceived as normal dog behavior. As a result, he had no appreciation of personal space. On top of this, the dog trailed her wherever she went, often walking at her heel from room to room. Their companionship, on the face of it, was rather sweet, particularly given that Sally had recently split up with her partner. But I knew it was exacerbating the problems that had developed.

When I asked Sally about her routine, it was soon clear there was none to speak of. Being a district nurse, her calls came at odd times during the day. There was no real consistency. She would usually leave in the morning but sometimes she would pop home for lunch, other times she would not get back until late at night. It was clear there was an element of guilt about this. For instance, the house was full of toys of every conceivable kind. There was also a bucket full of biscuits near the front door. When I asked what that was for, Sally explained that it was part of her leaving routine. As she left each morning, she would pat Bruce, tell him she'd see him later, then give him a biscuit as she left. The biscuits were left out there so that Bruce could help himself while she was away. There was no question that Bruce was very well loved but that Sally was channeling that love incorrectly. She needed to turn her affection in a different direction.

The diagnosis was not long in forming. I felt pretty sure that I had got a dog that was feeling responsible for its owner. Bruce felt that Sally was his baby and not vice versa, so that whenever she got up to move, he—like any good parent—followed to make sure she was all right. His attacks on the door frame were expressions of pure panic on his part. The area he had concentrated on was where the separation had taken place. His biting of the door was his attempt to break out of the house and reunite himself with his baby. When I explained to Sally what was going on, she understood his reaction perfectly. Wouldn't you be out of your mind with worry if your baby left you in that way? Apart from anything else it was all he could do. (It has now been proven scientifically that a dog's endorphin levels increase when it is chewing; like an adrenaline rush, chewing dulls the pain.)

In addition, Sally was doing a lot of things that did not help the situation much. To begin with, I pointed out that the way she was leaving the house was agitating Bruce. The ritual she went through before leaving each morning was underlining his status as the leader of their little pack. As he had begun to understand the ritual, he could anticipate what was going to happen. The dog felt that he was responsible and did not want her to go out into a world he felt she did not understand; an Alpha, by definition of its status, knows best.

His anxiety was heightened by her mood when she returned. Whenever she returned to find a mess, Sally scolded Bruce. As far as Bruce was concerned, this must be connected to something she encountered while she was out there. So the dog was anxious when she was gone and anxious when she came back because of what had happened to her. As if all this was not enough, the situation was exacerbated even more by the way she kept leaving biscuits at the door. Food is provided by the leader. So if you can get food at any time, you must be the leader.

Whenever I come across a case like this, I am reminded of that scene in *Peter Pan* when Wendy and the children fly off with Tinker Bell the fairy. As they leave, some of Tinker Bell's fairy-dust lands on their dog, Nana, who floats up with them. When her chain stops her floating any further, her face is wreathed in a mixture of sadness and terror. She is worried at where her family is going and desperate that she is not going along to offer her protection. I felt so sorry for that dog and I felt a similar sympathy for Bruce. Like so many of the dogs I come across, Bruce believed he was responsible for his owner. Because his origins lay in a society where the preservation of the pack was the key, the separation from his child was making him desperate. My job was to reverse his roles; his job description had to be changed.

Each owner I deal with has to begin the same way. The first thing Sally had to do was to go through the process of Amichien Bonding. Only by going through the four elements would she be able to rebalance the relationship so that Bruce was relieved of the responsibility that was causing him such stress. Sally's closeness to Bruce was such that she felt terribly guilty about ignoring him at first. Like so many people, she wondered whether her dog was being upset by this. To this day, people who begin the process say to me, "I'm sure my dog thinks I don't love it anymore." My response to this is that we are once more hung up on a human idea of the world, specifically our idea of love. If you really love someone or something, your whole motivation should be to do right by them. In circumstances like this, I ask owners to think less about what their need is and more about what the animal's need is. And besides, once the bonding process has been achieved, you are free to shower your dog with as much affection as you like—it is affection in a different direction.

Bruce was four years old, and he had been doing this for a long time, so he was definitely what I call a remedial dog. To deal with the specific problem of leaving the house, I needed to take the process deeper. The first thing I did was to get Sally to stop addressing the dog when she was leaving. I wanted her to behave like a leader and to come and go as she pleased. I also asked her to make a less drastic transformation in the atmosphere in the cottage when she left. While she was there the radio or television would be blaring away and she would be chatting to Bruce or on the telephone. The moment she walked out the door, all that noise disappeared. Bruce was left there in suffering silence. The house was going from this place where there was noise and activity to this place where there was nothing. It was obvious to the dog that she was about to leave.

I also asked her to stop leaving food around. The signal that was sending was completely wrong. It was underlining the dog's feeling that he was leader. Besides, it was a fairly pointless exercise. The dog wasn't eating the biscuits. What parent is going to sit down to a meal when it doesn't know where its child is? Instead, I got Sally to feed the dog herself, gesture eating so as to emphasize her leadership. I asked her to continue this for the next two weeks.

As far as I was concerned, however, the key thing Sally had to do was take the drama out of her departure and arrival, to make it seem like an ordinary occurrence. To help Bruce understand that Sally's comings and goings were commonplace, I asked her to try a technique I call "gesture leaving." I must admit Sally gave me a funny look when I first explained what I wanted her to do, but she went ahead nevertheless. I wanted her to leave without Bruce becoming agitated. She could not go out through the front door for obvious reasons. This was where all Bruce's anxieties were concentrated. Unfortunately, the cottage did not have any other doors so I asked her to leave by another exit, her living room window.

Before she did this, I got her to put on her shoes and coat in full view of Bruce. I also asked her to leave on the radio so that there was no discernible change of atmosphere. She then climbed up and out of the window, walked around the side of the house, and came back in through the front door. As she reappeared I made sure she completely ignored Bruce. The message she was sending out was that she was the leader and she would come and go as she pleased. She did not need to ask his permission to leave the house.

Sally thought the double-takes on Bruce's face were marvelous. He couldn't take in what was going on. More importantly, however, he wasn't frightened by what was going on either. Encouraged by this, I got her to repeat the process, this time staying out for five minutes. Once more, she returned to ignore Bruce. Once more, he was relaxed at the prospect of Sally having left and re-entered the house. On this occasion, as in the first, Sally returned to find the door untouched.

I am often asked why leadership needs to be reinforced every time you reunite with your dog. There are several answers to this. On the most fundamental level, the answer, once more, lies back in the wild. The make-up of packs is constantly changing. When a group of wolves heads off on the hunt, there is no guarantee each of them will return alive. There is always a possibility that the Alpha pair or their subordinates may be wounded or killed and fail to reappear. So after each separation, the hierarchy is re-established. The pack redefines its power structure so that at any given time it knows who is going to lead, who is going to defend the pack, and in what order the roles are to be carried out. It is an instinctive action on the part of the dog and it applies equally to the domestic situation. Whenever you leave the dog's sight, it has no means of knowing or understanding where you have gone or how long you have gone for. So whenever you reappear, regardless of how long you have been absent, your dog will need to know who is performing the role of leader. This is the only way the status quo can be maintained.

With this in mind, it was imperative that Sally keep doing this for an extended period. We began working over the weekend. I got her to stay out an extra five minutes each time she went out. By the end of the weekend, Bruce was noticeably more relaxed and the door had been left alone. I don't know what the neighbors were thinking of this woman jumping out of the window continuously but, frankly, neither Sally nor I cared.

Sally carried on doing the same thing when she went back to work. Soon, instead of rushing up to her when she reappeared in the evening, Bruce was standing there wagging his tail. The pair were more devoted to each other than ever. Soon Sally was finally able to make her long-awaited call to the local carpenter.

8

Mean and Moody: Dealing with Nervous Aggression

As news of my reputation with problem dogs spread, I was increasingly invited to take part in radio phone-in programs. Then, in the spring of 1999, I was invited by my local television station, Yorkshire TV, to apply my methods to six problem dogs. The dogs had been chosen from 600 letters submitted, and represented a cross section of the type of difficulties I am asked to sort out in my work. Among them was what appeared to be a bad-tempered golden cocker spaniel called Meg.

Her owners, Steve and Debbie, told me that she suffered from sudden mood swings; she would break into a high-pitched bark whenever strangers approached, she would rip up the letters when the postman came in the morning. Worst of all, she was a biter and had nipped the young daughter of a friend of theirs. Even the owners, who had three children of their own, admitted they were apprehensive of Meg when she was in "one of her moods." They confessed they had been advised to do one of two things: give her a sound beating or have her destroyed before she did any serious injury.

Even before I had met Meg, I was certain she was a classic example of a dog suffering from by far the most common problem I am asked to deal with: nervous aggression. Nervous aggression can manifest itself in an incredibly wide range of behavior. It is at the heart of the problems many, many people have with dogs biting, barking at, or jumping up onto visitors to their homes. It is the root cause of attacks on those most endangered of species: the postman, the milkman, and the newspaper delivery boy. Yet for all its myriad manifestations, nervous aggression can be conquered by making one fundamental change: removing from the dog its status as leader of its pack.

No dog decides it is the leader of its pack of its own volition. The dog knows instinctively that there must be a leader for the pack to survive, and Meg's owners had inadvertently given her the position by the signals they had given out. Given this, Meg's behavior was perfectly understandable. She was merely trying to carry out the job she had been given. Her aggression was down to the fact that she had been placed in a situation where she had no experience and no guidance and was operating in a world of which she had no understanding. Her ferocious attitude toward strangers was her way of repelling intruders she believed might pose a threat to her "pack." As if to exacerbate all this, Meg was the only dog in the household. Ask any lone parent about the stress that role brings with it!

As Steve and Debbie were discovering, in this situation an owner is powerless to help. Indeed what he or she sees as assistance is usually the opposite. The dog does not look to the owner for advice. In its mind, if the owner was superior, stronger, and more experienced, he or she would be the leader. So the owner is ignored and, if too persistent, reminded of his or her subordinate role via aggression. It was little wonder the whole household was becoming wary of Meg and her moods.

I understood the feelings Meg's owners were suffering all too well. They loved their dog, and they just wanted to help it. What they did not realize was that the best way to provide that help was to let it know who was boss. By doing that they could give it some peace and take the pressure off it.

In all the work I do, I like to lead by example. If owners are to follow my method properly, I need to show them precisely what can be achieved by establishing leadership. So from the moment I first walked into the living room I refused to acknowledge Meg in any way; there was no eye contact, no touching, nothing. As well as underlining my Alpha status, this conveyed to Meg that I posed no threat to her or her wards. To underscore my status further still, I also ensured I looked as if I belonged there, in fact as if I owned the place. People are constantly amazed at the power contained within this simple act. Rather than making her customary fuss, Meg simply ignored me in return. Even this was a revelation to a family whose normal reaction by now was to panic whenever a newcomer came into contact with Meg.

My challenge now was to enable Meg's owners to behave in the same authoritative manner. So the first thing I did was to get Steve and Debbie to leave the room without acknowledging her. I then asked them to return to the room continuing to ignore Meg's behavior, whatever it was. Like most owners, they found it an unnatural thing to do at first. It was a step into the unknown. They had seen their dog display such eccentric behavior, part of them, I'm sure, feared how she would react to this sudden snub. But the more I explained to them, the more they understood their constant deferring to Meg was prolonging her reign of terror. Each time they acknowledged her—in whatever way—they were re-establishing her position as leader. And by doing that, nothing was going to change.

Like so many of my clients, Steve and Debbie were genuinely determined to address the problem and agreed to go ahead as I explained. Meg, of course, was agitated in the extreme. Her eyes were glaring at me, almost out on stalks. She was pacing up and down, letting out this low grumble; she was perceptibly trembling. When she had calmed down a little, I then got Steve and Debbie to begin calling Meg to them, using small pieces of dried liver to reward her compliance. Within an hour, her owners were sitting alongside a dog that was tangibly less stressed than she had ever been before. Most importantly of all, the glowering eyes had been replaced by what I like to call "soft eyes." In the years I have been using my method, I have come to recognize soft eyes as the clearest signal that a connection has been made, that I have communicated with the dog. As soon as I saw Meg's eyes, I knew a corner had been turned.

I continued working with Steve and Debbie for two weeks, making sure they kept asserting their leadership during this time. They grasped the principles of Amichien Bonding well. They would ignore Meg whenever she approached them uninvited. All attempts by her to establish contact were disregarded. Whenever she made a positive response, she was rewarded with a morsel of food.

At the same time, I concentrated on teaching them to react differently whenever Meg became agitated. If she barked at the postman, one of the family would acknowledge the bark with a simple "thank you." The message was that Meg had done her job, passed on the information to the newly elected leader.

Old habits die hard in dogs as well as humans. For a while, she continued to growl at visitors when they came into the sitting room. Whenever this happened, I asked Steve and Debbie to get up and walk out of the room. This simple action made two powerful points clear to Meg. Firstly, there were consequences to her actions. Secondly, it was no longer her role to decide who was and was not welcome in the home. Her days as leader were over.

Finally during this period I got the whole family to gesture eat. Each of them would make a point of eating a cracker or a biscuit in full view of the dog. Only when each of them had finished was the dog's bowl put down on the floor. Her owners were signaling: "Right, we've finished, you have what's left." It was, as I have explained earlier, another important way of underlining the pecking order and relieving the dog of its responsibility for a job it was not equipped to do.

Within a few weeks, Meg's personality—and the atmosphere within the entire family—had been transformed. The arrival of the morning post was no longer a cause for consternation. If Meg showed any signs of agitation, a few simple reassuring words from her owners calmed her down. The days of the headlong dash to the doormat were over. Visitors were also free to come and go without being molested or intimidated.

The idea behind the television program was that viewers would see the dog "before and after" a period applying my method. With the cameras still running, Steve and Debbie confessed they were overwhelmed by the transformation that had occurred. They could not hide their emotions as they cuddled Meg in a way they had never imagined possible. Debbie cried openly. Moments like these make what I do worthwhile. Sitting there in the room with them I found it impossible not to shed a tear or two myself.

9

Peacemaking: Dogs That Bite

The most dangerous, disturbing, and difficult problem I have to deal with is undoubtedly biting. I only have to cast my mind back to my own Purdey to recall the chilling awfulness that comes with the realization that your dog is capable of attacking a human. To most people, as it did to my father, biting represents the crossing of a line, a step into a form of behavior that is simply unacceptable. I have lost count of the number of times I have been asked to intervene in cases where dogs have been given their last chance to reform or face being destroyed. I have been lucky enough to save most of them.

In dealing with this subject, we have first of all to be realistic. The plain truth of the matter is, of course, that dogs cannot unlearn what they have been programmed instinctively to do. Their right to self-defense is as deeply ingrained as ours. Placed in a threatening situation, they face the three options—to flee, freeze, or fight. Make no mistake, if necessary they will exercise the latter, and will take up their right to defend themselves. It is as simple as that.

As with all areas of my work, no two biting dogs are alike. The root causes underlying their behavior may be the same, but the manner in which their aggression shows itself is, by nature, unique in every case. This was certainly true in the cases of three very different dogs I have been asked to treat since developing my method.

* * *

Years of experience have taught me to recognize certain types of dogs without even casting eyes on them. Such a dog was Spike, a white German shepherd, owned by two brothers, Steve and Paul, living in a suburb of Manchester. The brothers had called me in the hope I could cure Spike's habit of attacking and biting visitors to their house. Spike's attacks had become increasingly forceful. He had, for instance, begun attacking anyone who tried to leave the house. The moment anyone, the brothers included, placed their hand on the front door handle, Spike would jump up and deliver a nasty nip. Family members had become so fearful of the situation, they no longer visited. Steve and Paul were seriously worried they would have to let Spike go if the situation did not improve.

I didn't even need to enter their home to realize Spike was a formidable creature. From the depth, tone, and power of the bark and the furious speed in which it was delivered as I walked up the path toward the brothers' front door, I could tell this was a dog that was supremely confident in itself and its status within its pack.

It was an impression that was confirmed as soon as I was inside the house. Within the safety of his own den, Spike simply oozed authority. There was an almost tangible aura around the dog. As he strutted around, his body language was unmistakable. He was a well-built animal and was aware of his power. He was the Alpha male in this home and was determined that everyone would know it. As I walked in, he eyeballed me, barking and threatening about three feet away from me.

As I have mentioned before, respect is absolutely central to all relationships with dogs. If you demonstrate it to them, they will reciprocate by showing it to you too. In the case of Spike I knew that it would be especially important. As ever, my first job was to convince Spike that I too was an Alpha. In this case I also had to persuade him that I was a non-threatening one. I began by immediately ignoring the dog in the usual way. On this occasion, however, I was also careful to avoid any sudden movements that would be likely to cause Spike anxiety. Again, experience has taught me that even the most innocuous movement, crossing our legs for instance, can provoke a response in a strong-natured and aggressive dog like this. It was a thin line to walk in many ways; I could not be seen to be weak, but at the same time I could not throw out signs of hostility. At the back of my mind, as always, was the model of the wolf pack. My aim was to create a situation in which we respected each other's mutual space.

The brothers had asked the advice of many people before turning to me. I was appalled at some of the things they had been told. They had been advised, for instance, that the dog needed a good thrashing. "Knock some respect into it," the so-called expert had told them. Another person had—to my horror—recommended that they simply "stare out" the dog. Short of physically attacking the dog, I can think of nothing more certain to cause a confrontation. This constitutes a direct challenge to the dog, and in the case of dogs like Spike, they will invariably defend themselves. Fortunately, the brothers were too sensible to undertake either course of action. I shuddered to think what the consequences might have been if they had not been so wise.

As soon as I began explaining the situation that existed, Steve and Paul began to see hope at last. Spike clearly saw both the brothers and the house as his responsibility. His aggressive behavior at the door in particular was clearly to do with his protecting the den. He could not rationalize what lay outside the door but he was certain that he was responsible for guarding his pack from whatever dangers lay out there. Talking to the brothers in more detail, I learned that Spike's reaction was in fact a nip rather than a bite, which did not surprise me. Very few dogs bite to cause damage. What they are doing is delivering a warning shot. If a dog, especially a German shepherd like Spike, really meant to bite you then it would; the carnage it would cause does not bear thinking about.

Spike's protectiveness was, in truth, typical of dogs in the pastoral group, like collies and shelties. They have been bred by man to do the job of caring, and that is what they do to the best of their ability in an environment they don't understand. As I got to know Spike and his owners better, it became clear that his increasingly aggressive behavior was being made worse by the fact that everyone deferred totally to him within that home. As his leadership had remained unchallenged, so his power base had grown. This situation had to be reversed; the owners had to exercise what I call "power management."

My aim was to allow the brothers to establish a first foothold of their own within the power structure of their pack. To do this, I had to help them create as calm and non-threatening an environment as possible. Fortunately, I found a hugely helpful ally in the unlikely shape of their housekeeper. Some people undoubtedly have a greater confidence with dogs than others. Sometimes I wonder whether these people are somehow in closer contact with the ancient language that has been lost. Equally, however, there are others who are awfully nervous around dogs. We all know people who tiptoe around, almost creeping along the walls it seems, whenever they come into contact with a dog. Their nervousness is, of course, picked up instantly by the dog. The fact is, however, that no one need harbor fear of any kind. Treated correctly, almost all dogs are perfectly safe and will cause no harm to anyone.

There was no doubt into which category the brothers' housekeeper fitted. She had been present in the house throughout my session quietly getting on with her cleaning, washing, and polishing duties. She barely paid the dog any attention whatsoever. In return Spike treated her with the utmost respect. At one point he even jumped out of the way when she appeared with her cleaning trolley.

I was able to use her as a means of explaining what the brothers needed to do. They could see that there was nothing intimidating about the lady at all. Yet by instinctively refusing to pay homage to the dog she had persuaded Spike that she was his superior. If there was a model of the behavior they should be aiming for, she represented it perfectly.

I knew the challenge that lay ahead of the brothers was immense. I told them that on a scale of one to ten in terms of aggression, Spike was easily an eight—way above the average mark of four or five that I was used to dealing with. I warned them that they might need to keep up the silent pressure for months rather than the usual weeks. Again, fortunately they were willing students and they adopted my methods enthusiastically. They called me occasionally over the following fortnight, usually to check on dealing with specific situations. In most cases they were doing exactly the right things; they had grasped my ideas perfectly.

Four months after I had visited them, I got a phone call from a relative asking me to help with a problem he was encountering with his dog. He told me that Spike's behavior had improved enormously. The brothers were now able to control any situation that occurred within the home. Their family had begun visiting them again.

* * *

Not all dogs exude the same blend of confidence and power as Spike, of course. It does not make their aggression any less dangerous, however. In November 1996, I began a BBC radio phone-in program offering help with problem dogs. One of my earliest callers was a couple called Jen and Steve from the town of Driffield, a forty-mile drive away from me. They had taken on a little three-year-old cocker spaniel called Jazzie six months earlier. He had a reputation for bad behavior but, as previous dog owners, they had been confident they could improve his temperament. Their efforts had, however, failed. Even worse, Jazzie had begun biting both of them whenever he disapproved of what they were asking him to do.

Once more, a clear idea of the dog I was about to deal with was forming even before I met Jazzie. Walking up to their front door, I heard furious barking, but this time of a kind much different from that of the super-confident Spike. This was a very staccato, almost panicky bark. My suspicions were confirmed as I stepped inside. As I was greeted by Jen and Steve, Jazzie pushed himself forward barking more aggressively now. His body language was as confrontational as it could have been but the crucial difference was in his position. Unlike Spike, who had been "in my face," Jazzie was standing at least six feet away from me. To me it was clear in an instant that this was a dog that was even more terrified of the situation than the humans with which it was coming into contact. It was clearly a reluctant Alpha, a dog that had been given the job of leadership but was completely unsuited to the role. Once more, we had to strip him of his responsibility.

As I have explained, all dogs react to the signals I present to them at their own pace and in their own way. Some dogs, like Spike, are particularly reluctant to give up their responsibility. Their self-belief is such that they cannot face the prospect of losing their top-dog status. We see this in our politicians, of course. Look at how people like Margaret Thatcher cling to the idea that they remain in power even when they have little left. Other dogs, however, are utterly relieved to have the load lifted off their shoulders. Jazzie was an example of the latter.

I began working with Jen and Steve in the normal way, explaining my method and getting them to put it into practice immediately. Throughout the time we talked, Jazzie stood in the same room, again keeping his distance slightly but maintaining an almost continuous stream of barks and growls. I, of course, am immune to such interruptions but, as so often happens, the owners reached the breaking point and asked whether they should remove him. I told them to try to ignore it, which they did. Within half an hour their persistence paid off. Suddenly Jazzie stopped making a noise, turned away from us, and headed toward a staircase in the middle of the open-plan room. In full view of us, he walked to the top of the stairs where he promptly plonked himself down and sat with his back to us. If it had been a child, we would have concluded that he had gone off in a sulk.

In all situations, it is vital that the dog is allowed the flight option, that it is free to walk away from the situation. The worst possible thing we can do is to put a dog in a corner. That way the next two options—to freeze or to fight—are brought into play. And that is where the real problems begin. For this reason we let Jazzie sit there. Jen and Steve wondered whether we should pick him up, but I reassured them that Jazzie was doing exactly what was required of him. It was as clear an example as I had seen of a dog dealing with the new situation and making a judgment about its future. I advised that Jen and Steve should not in the future go up to Jazzie but should invite the dog to come to them. This is vital with reformed biters; such dogs must not be put into a position where the only defense is attack.

Jazzie sat on the stairs for at least half an hour. Then suddenly he picked himself up again, trotted back down the stairs, and came to lie on the carpet. Soon he was stretching out on the carpet in front of us. I recall the sunshine pouring into the living room. And I couldn't help thinking that the shadows were lifting from Jen and Steve's life too. The balance of power had shifted perceptibly in that one hour. Suddenly it was as if Jazzie didn't have a care in the world. He no longer felt responsible for anyone in the room. Instead he was now waiting for the opportunity to pay homage to his new leaders. Jen and Steve could begin to enjoy a new and fulfilling life with him. It was only later that I realized Jazzie had been within a few days of being destroyed. My intervention had been the final throw of the dice. The sense of job satisfaction was enormous.

As a footnote to this, I should mention that two years later I got a call from Jen. She and Steve had become concerned at the fact that Jazzie had begun growling and barking at visitors once more. He had also nipped them when they had tried to take objects away from him. When I asked her whether they were still adhering to the five-minute rule, she admitted they were not. Jazzie's behavior had improved so much that they had, if the truth be known, become a little bit blasé about it all.

I told Jen what I tell all the owners I come into contact with. My method is a way of life, not a quick fix. It must be adhered to at all times, and must become second nature. What was particularly pleasing in this instance, however, was the speed with which Jen and Steve were able to rectify the situation. I advised them to go back to the beginning, to "shut down" on Jazzie again, as they had done at the beginning of the process two years earlier. I always form an interest in the families I help. So I rang Jen the next day to find out how things were. She simply laughed. Jazzie was back to his best behavior once more, she told me. It had taken four hours applying the method to iron out all his problems.

* * *

Of course whenever I treat a case involving a biting dog, I cannot help but think of Purdey. Each time my mind goes back to the awful events of almost thirty years ago.

Purdey's behavior, I now know, was typical of so many dogs. She was no different from Jazzie and Spike; she was merely trying to do the job she believed she was supposed to. It was not her fault that she was completely unequipped to fulfill that task. When Purdey had jumped up and barked at my son Tony, she had been treating him as a subordinate pack member. He had been inadvertently challenging her leadership, and she had dealt with him in the manner she believed was right. It was her misfortune that when she did so, he was standing in such a dangerous position.

If I had my time all over again, I would have reacted completely differently to the behavior that led up to that moment. I would not have chastised her when she misbehaved; I would have understood that when she ran off into the countryside, she believed she was heading off on a hunt, on a mission to help me and my fellow pack members. Blessed with the knowledge I have now, I would have relieved her of the responsibility of leadership and allowed her to lead a less stressful existence long before we got to that fateful moment. Hindsight, of course, is a wonderful thing, but it will not change what happened to Purdey. It does, however, provide me with the inspiration to do all I can to save every Purdey I come across. And that inspiration is never stronger than in cases where children are involved.

There is no doubt in my mind that dogs do view children differently to adults. I believe there are two reasons for this, the first being that dogs find children even more confusing than grown-ups. If we think about it, a dog must find children particularly baffling. They speak faster, move faster, and behave far less predictably than adults. As I have explained, calmness and consistency are essential in establishing a relationship with a dog. These are hardly words normally associated with children.

The second reason is even more obvious. Children are, in

the most literal sense, nearer to a dog's level. For this reason the animal tends either to see them as a threat or as creatures deserving of extra protection. The former, of course, is a difficulty many owners find hard to deal with. My own view is clear: very small children and dogs should be separated wherever possible, or supervised. Both need space in which to develop and should be given that space.

The sight of a dog protecting a child, on the other hand, represents something rather lovely to behold. I don't believe there is a more magical bond. It is an incredibly powerful link, one I had seen myself years earlier with my own dog Donna. Even here, however, the bond can bring problems, as I discovered when I was called to deal with Ben, a commanding black mongrel who lived with his owners Carol and John and their nine-year-old son, Danny, in Salford, Lancashire.

Ben clearly doted on Danny and had become fiercely protective of him. His most aggressive behavior had been directed toward John's father, Danny's grandfather. It was not hard to see why, of course. Grandad lived a hundred miles or so away in Wales and saw the family infrequently. Whenever he arrived at the house, he would shower the boy with affection. Ben had no concept of the relationship that underpinned this; he simply saw the senior member of the family as a threat and had begun to physically attack the grandfather. The situation had become so bad that grandad would sometimes be confined to an armchair, unable to make a move without Ben growling and glaring at him menacingly.

The strain such situations can place on a family is immense. Loyalties become confused. Owners are accused of caring more for their animals than their own flesh and blood. It can cause enormous damage. Again, luckily I was dealing with a family mature enough to deal with the problem. I began by tackling the situation in the normal way with the adults. They took to the process of Amichien Bonding well enough. But I knew that involving Danny in this was the key to success.

Involving children can be one of the most difficult elements of the process I practice. Quite understandably many fail to understand what is being attempted. As I have explained, in the case of very young children I recommend segregating them from dogs if they become too boisterous together. By the time children are three or four, however, they are capable of understanding much of what is going on and will be able to participate in the process, particularly if it is presented as a game. In my experience, teaching a young child to ignore a dog if it comes at them can work rather well. That said, even if it is presented as a game, they can grow bored, as they do with most other games; so in the end it is a matter of personal judgment on the part of parents.

In the case of Danny, however, I had no hesitation in involving him in the process. Apart from anything else, in this case his help was going to be vital in dealing with Ben. Understandably, Danny found it very hard indeed to stop stroking Ben. When I asked him to stop doing this, he told me he found the job of ignoring his playmate incredibly hard. But, with his parents' permission of course, I explained the potential consequences. I gently told him that if we did not get this right, Ben might not be his friend for much longer. I was not trying to scare the boy, I was simply trying to get the message across. Fortunately it worked, and for the rest of the session Danny dealt with the situation by stuffing his hands in his pocket whenever he was around Ben.

The session lasted two hours, during which time Ben did all he could to get the family's attention. By the end of that time, everyone's nerves were beginning to fray I must confess. It was at this time that Ben showed them the value of what they were doing. By now, Ben had exhausted his attention-seeking repertoire and lay down in a favorite spot in front of the fire. When I saw this I knew he had realized his time and energies were being wasted. With the atmosphere more relaxed, Grandad got out of his chair and walked across the room. Without thinking as he passed his grandson he automatically placed his hands on the boy's shoulders. Ben remained on the fireside carpet, unperturbed. By the end of my visit, the tension surrounding Ben had visibly lifted. When I spoke to the family again a few weeks later, they proudly told me there had been no more confrontations. Danny was now looking forward to more frequent visits from his grandfather.

10

The Bodyguards: Overprotective Dogs

The dog's reputation as man's best friend is a well-earned one. In addition to providing entertainment and companionship, a dog's caring nature and sheer physical presence give many people an important sense of security, and we have all seen even the meekest of animals transformed into aggressive devils when their beloved owner is threatened.

A pet's protective streak is not, however, always a force for good, especially when it is applied within families. I have dealt with a number of cases where favoritism toward certain family members has caused consternation. The most extreme example I've come across was that of Toby, a springer who lived with a couple, Jim and Debbie, in Grimsby. Toby's protectiveness of Debbie expressed itself at night-time. As a result, she and Jim had grown to dread going to bed.

By day, Toby was a reasonably well-adjusted dog, but at the end of the evening, however, he transformed. The moment Jim and Debbie began switching off the lights in the house and heading upstairs, Toby would bound up the steps ahead of them, rush into their room, and jump onto their bed. While he would allow Debbie to climb in without protest, Toby would snarl and growl at Jim the moment he made a move for the duvet. His determination to keep husband and wife apart was so great that Jim genuinely feared he would be bitten.

Jim had resorted to all sorts of tactics to beat Toby into bed, everything from sneaking up ahead of Debbie to distracting the dog by pretending there was some kind of danger in the house. Jim would walk off into another part of the house and begin banging loudly on a door, then the second Toby went to see what the trouble was, he would rush back into his bedroom and dive under the duvet. At first their predicament seemed funny, but by the time Jim and Debbie called me, it was far from a joke.

Few facets of a dog's behavior are as fundamental as the protectiveness that Toby was displaying; in effect, he was behaving like a jealous spouse repelling a rival suitor. It is, on the face of it, a difficult point to grasp, yet the principle becomes clear when we consider the pack environment. As I have explained, the rules of life within the wolf pack are founded on the primacy of the Alpha pair. To the dog's ancient ancestor, these two dogs rule supreme, and their status is so unassailable that they are the only animals allowed to reproduce. The key thing to realize here is that, as the lone dog in his "pack," Toby looked to his human subordinates for a partner. And his choice had been Debbie rather than Jim. The prospect of them being intimate was utterly unthinkable to Toby, and the idea of Jim (in Toby's eyes a subordinate pack member) sharing a bed with Debbie, his Alpha partner, was a threat to everything that held his world together. When one looks at it from this point of view, it was little wonder Toby was so protective. His instincts would have told him Jim and Debbie were male and female, only adding to his anxiety.

It often takes owners some time to accept the diagnosis I give them. This was certainly the case with Jim and Debbie who found it extremely hard to accept that, in effect, Toby was acting like a jealous boyfriend, repelling the advances of a rival suitor. As I talked to them and as they began to apply my method, however, they soon came around to my way of thinking. The first thing I asked them to do was to keep Toby out of their bedroom. Personally speaking, I have no problem with dogs sleeping in owners' bedrooms. I would not go so far as to let them sleep on the bed; otherwise, however, I see no harm in them sharing the room if it makes people happy.

If Jim and Debbie found Toby had slipped into the bedroom without their knowing it, I asked them to use the reward principle to draw him out of the room. If he jumped up while Jim was in bed, Jim was to wriggle around and make the dog as uncomfortable as possible. The important thing I stressed was that the dog should never be forced off the bed. Any confrontation would force the dog to consider fighting—and that was a situation no one wanted. Far better to engineer the situation so that the dog was removed from having to make that choice. Toby's behavior soon improved, and the end of Jim and Debbie's day was soon a more relaxed and enjoyable time.

* * *

Being the amazingly intelligent creatures they are, dogs have evolved a huge repertoire of tricks to assert their authority, and dogs like Toby exemplify just one common method. I have also encountered a number of dogs that have a habit of leaning lightly against their owners. This can often build up so that the force the dog is applying effectively blocks any forward movement. It is a clever trick.

It is easy to see what is going on in this situation. The dog is trying to direct the owner's movements, impose its will, and, once again, establish the fact that it is in charge. This, of course, is bad practice and cannot be allowed to continue. If I am to be honest, this was not a habit I had seen very often until I began taking my work out into the wider world. Since doing so, though, I have seen numerous cases, the most memorable being a German shepherd called Zack.

Zack's owner, a lady called Susie, loved sitting on the floor with her pet. Of course, in most normal circumstances there is nothing nicer or more natural than being able to sit quietly with one's best friend like this. The problem was that Zack took leaning to the extreme. Whenever Susie sat next to him, Zack would not just lean on her, he would place himself across her legs so that she was pinned to the spot. I saw it for myself when I visited them. The moment Susie sat down, Zack tipped his body into hers. Susie's knees were tucked up at first, but Zack literally forced her to flatten them out onto the floor. He then spread himself right across her legs. German shepherds are big, powerful dogs and Susie was a relatively slight lady. She was, to all intents and purposes, Zack's prisoner; she wasn't going anywhere without his permission. As if to underline his status yet further, Zack then positioned his stomach so that Susie began tickling him. This too, it turned out, was a regular part of daily life in the house.

Zack was clearly manipulating Susie into adopting a routine he had chosen. As they sat on the floor together, the first thing I asked Susie to do was to stop tickling him. She was reluctant to do this. "He'll get upset and start growling," she told me. Sure enough, the moment she stopped the tickling he began to make rumbling noises. She understood what was required of her, however, and went on to free herself from his body lock. She just took her legs from under the dog, got up, and walked away. From there she began the basics of Amichien Bonding, taking specific care in this instance to remove herself whenever Zack tried to impose himself physically on her. Each time he did so, she broke loose. Zack was soon learning the consequences of his actions, and Susie was soon able to lie on the floor beside him.

* * *

All of us, I'm sure, have encountered homes that are overseen by an overprotective dog. At the first sight, sound, or smell of a passer-by, the dog comes belting out, barking and bouncing around in the most animated way it can manage, pacing or even racing up and down the perimeter wall or fence of its domain as it does so. The message it is sending out is clear: you are dangerously close to my territory, stay away for your own good. Many people do exactly that.

Such behavior, particularly from a loud and aggressive large breed of dog, can be a serious source of distress for passers-by. It is common for people to make a point of crossing the road or even making a detour so as to avoid confrontations. Children, in particular, can be terrorized by such dogs. Of course, there are an unfortunate few owners who revel in their dog's aggressive reputation. Equally, there are an unpleasant few passers-by who will deliberately inflame these animals, winding them up into an even greater frenzy in order to satisfy their warped sense of humor.

In most cases, however, the fact of the matter is that this behavior is as upsetting for the owner and the dog as it is for the passer-by. The root cause of this problem, what I call "boundary running," is of course territorial. The dog believes it is the leader of its pack and sees all approaches to the perimeter of its den as a potential attack on his domain. In the course of my time treating dogs, I have seen dogs that have become deeply distressed by the burden of this responsibility. One case springs to mind in which the dog would run around the boundary of its owner's circular garden. The poor animal would run and run—in ever-decreasing circles yet in an ever-increasing state of anxiety. The good news, as two case histories will hopefully illustrate, is that boundary running is a relatively straightforward problem to treat.

The first case involved a lady called Mary and her Border collie, Tess. Mary and Tess lived in a home on the corner of a housing estate and, as a result, had an almost constant stream of pedestrians walking along the perimeter of their garden. The main problem, however, lay with one particular neighbor, another lady who walked her dog, another Border collie, past Tess and Mary's home at the same time each morning. The sight of this other dog would set Tess off every time. She would run along the perimeter of the fence barking and snarling as she did so. Truth be told, the other owner seemed to encourage the dog to fight fire with fire. It too would leap around aggressively, agitating Tess even more. Mary had tried her best to overcome the problem, but with no luck. By the time she called me in, she was at the end of her tether.

Mary had made all the most common mistakes. She had, for instance, got into the habit of shouting at Tess. Owners who say "stop it" are only guaranteeing their dog will do the complete opposite and continue. By doing so, they are acknowledging what the dog is doing and winding the dog up even more. I began by asking Mary to go back to basics and begin the Amichien Bonding process. In addition, I asked her to keep Tess in the house for a day or so while she got the message across. I felt that by connecting with Tess in this way, she would be in a much stronger position to get the right message across when the time came.

The test came a few days later, when Mary let Tess out in the morning. Tess's old adversary turned up at its usual time, and as usual, Tess responded to the challenge by barking and running along the perimeter wall. Mary's task was to relieve her of the responsibility of patrolling the boundaries of their "den." To do this, I asked Mary to extend the request and reward principles she had been applying inside the house. Tess was in such a state, she barely noticed Mary walking up to her. Knowing this would happen, I got Mary to touch Tess's collar lightly to attract her attention and then offer her a treat. In cases like this, in which owners are dealing with deep-seated, highly distressing situations, I ask them to give their dogs treats that reflect the special nature of the circumstances. It is, of course, the owner's choice what they use. I personally choose cheese, which my dogs love but only get on rare occasions. The special treat underlines the message that certain actions bring certain, pleasant, consequences.

Mary used her treat to gain Tess's attention. Once she had done so, she used the skills she had learned to lead her inside away from the situation. She did the same thing the next day, again gently encouraging Tess to walk away from the situation. This was no quick-fix situation; it would take time. Mary persevered and, by the fourth day, Tess's anxiety had reduced to such an extent that she would notice Mary's approach before she got to the fence. Soon Mary only had to walk three quarters of the way to the fence before Tess came to her for her treat. Tess was clearly getting the message.

After a week, this had progressed even further so that Mary only had to stand on the doorstep, fifty feet away. Tess was still barking at the other dog but nowhere near as intensely or as furiously. When she saw Mary at the doorstep, however, she returned to the house and the situation calmed. After another few days, she would not even go to the fence. Her mild barking would go on halfway down the garden. Eventually Tess—and the other dog—got on with their lives in peace. The morning ritual was no more.

* * *

I have been asked to deal with quite a few boundary runners in recent years. In the case of a pair of schnauzers called Kathy and Susie, my job was to treat two dogs at the same time. Because of the unusual set-up of their home, Kathy and Susie had a huge boundary to guard. Their house was set at the rear of a row of twenty or so terraced houses. This meant that all their neighbors' gardens backed onto the large grounds at the front of Kathy and Susie's home. At the first hint of any of these neighbors entering his or her garden, Kathy and Susie would spring into action. Understandably, the neighbors were not very happy about this. The owners were unhappy too; they didn't want their dogs being a nuisance either.

I remember visiting them on a warm summer's evening. To be honest, they had their doubts about whether my method could work. Fortunately, however, Kathy and Susie helped me prove my point soon enough. The fact that there were two dogs involved made little difference to me. From the moment I arrived in their home, I established my leadership credentials by giving out the simple, powerful signals that I always use. An hour or so after my arrival, the two dogs heard someone in one of the neighboring gardens and bolted off to defend their fence. I let them go. Then, without shouting, I went to the front door and asked them to come. The owners watched, slack-jawed, as the dogs turned and ran straight back to me and the reward I had waiting for them. Needless to say, the owners took the method seriously from that moment on.

The transformation was not going to be as dramatic as it had been when I called them to me, of course. It takes time for an owner to realign his or her relationship with their dog. And they are not going to see results until the bonding process has succeeded and the dog has fallen into line. It is a question of consistency and patience. In this case, I also asked Kathy and Susie's owners to approach the neighbors for their help. While they attempted to apply my ideas, they asked the neighbors to ignore the dogs completely. They were fortunate in having a group of understanding neighbors, and soon both they and the owners were being rewarded. Slowly but surely, the dogs were drawn away from their boundary confrontations. Within a week, Kathy and Susie were oblivious to the comings and goings next door. As the rest of that lovely summer wore on, both they and their neighbors got on with enjoying their gardens in peace.

11

The Up-and-Down Game: Dogs That Jump Up

Some owners cope with dogs jumping up at them; some even find it endearing (these are usually the owners of small dogs!). However, in many cases, it makes coming home an ordeal; muddy paw prints all over clothes and groceries scattered on the floor are just two of the results jumping can cause. The worst aspect of this problem for me is the lack of understanding between dog and owner; neither understands what the other is trying to tell them, which is where I can help as a translator, if you will.

All the dogs I deal with are memorable in their own way but none of the pets I have dealt with is quite as unforgettable as Simmy, a fawn, mixed-pedigree cross between a whippet and a terrier. Simmy's owners, a couple, Alan and Kathy, from Scunthorpe in Lincolnshire, called me at their wits' end. They told me that Simmy's particular problem was jumping up at people who came to visit their house. I knew jumping up was a particularly annoying habit in dogs. It took my first encounter with Simmy to illustrate quite how irritating it could be.

No sooner had I walked into the owners' home than Simmy was bouncing on his hind legs doing everything he possibly could to reach my eye level. I had seen this in many, many dogs before, of course. What marked Simmy out as different was the sheer athleticism he displayed. He was no more than fourteen inches in height but he could easily spring up four feet off the ground, trying to reach my eye level. What was even more impressive was the fact that he could keep doing it nonstop. (He was a perfect example of a crossbreed in that respect, blending the elasticity of the whippet with the sheer persistence of the terrier.) He reminded me of Tigger in the *Winnie the Pooh* books. Like Tigger, bouncing was what Simmy did best. His owners told me that he did this with every stranger that visited them and would persist whether they were standing or sitting. Needless to say, it was both embarrassing and uncomfortable. The owners were visibly on edge. I knew I would have a hard job on my hands.

As I have explained, body language is one of the most powerful means of communication available to dogs. And there is no clearer physical signal than that which lies behind jumping up. Again, we need to go back and look at the behavior of dogs and wolves in the wild to understand the principles at work. Dogs use their physical presence to establish superiority. It is, of course, a trait we humans have too. If you don't believe me, watch the body language of two boxers squaring up to each other at the beginning of a bout. Both are looking to establish some form of psychological advantage even before the physical warfare begins. Both are looking to send a clear message: I am in charge here, and I am about to show you who is the boss.

In the case of wolves, however, it is more than mere posturing. And it begins at a very early age. A form of jumping up exists among puppy litters. As they indulge in the routine rough and tumble that all young animals go through, wolf puppies place the top half of their bodies over the all-important head, neck, and shoulder area of their litter mates. Placing themselves in this position establishes an important signal that is repeated throughout their life in a pack. And again, it is all to do with status and leadership.

Within the adult population of the pack, the Alpha pair use this physical domination to reaffirm their primacy. They go through a similar routine on returning to their pack from the hunt. By lifting themselves above their fellow pack members and arching themselves over the same crucial area of the body—the neck and head area—they are simultaneously demonstrating their affection for their subordinates and reminding them of their ultimate power. The message is plain: I know how to subjugate and if necessary kill you. Acknowledge my leadership once more.

To overcome Simmy's problems, I would have to present him with some equally powerful language. In most cases, jumping up is one of the easiest of habits to overcome using my method. The key is to remember not to engage the behavior in any way. So as the dog jumps up in front of you, simply step back then move away from it. If you are restricted by space or the dog is overly excited, simply use your hand to parry the dog and gently push it away. Make sure in both instances you do not speak to or make eye contact with the dog. Remember you must not pay homage to its leadership.

As I have explained, Simmy's incredible exuberance took even me by surprise. I did not, however, allow it to deter me from my normal opening gambit. As I walked into the home, I studiously ignored him. This took some doing I have to say. At times, he was jumping up and placing his face directly into mine. At this point, I remember Alan became understandably upset. He reached out to grab Simmy's collar, clearly determined to bring his dog to earth through force. I insisted, however, that he didn't do this. The key, as always, was that I wanted the dog to exercise self-control. I wanted it to do things of its own free will, not to have its owner's will forced upon it. I am sure it was a difficult urge to resist but he agreed. As Simmy continued bouncing up and down in front of me, I simply talked over (and sometimes around!) him, outlining the process I wanted Alan and Kathy to undergo. In a nutshell, I didn't want them to join in the up-and-down game that Simmy was playing. Every time Simmy jumped up, the owners were responding. Every time they did so, they were acknowledging him. It had to stop.

I carried on talking to Alan and Kathy as we walked into the living room. As we did so, Simmy walked backward in front of me, still bouncing away as he did so. It was an Oscar-winning performance if ever I saw one. And it was precisely what I wanted him to do. It was not long before his behavior was changing, however. The cleverest dogs are always the most difficult to realign. They are constantly asking the question why? Why should I do what you say? Why can't I carry on doing what I please? Simmy certainly fell into the smart cookie category. So when he discovered his behavior was not getting any reaction, he switched tack and began barking at me loudly. Again, his poor owners were beside themselves with worry at the situation. But again, I just ignored what was going on and refused to engage Simmy. At the same time, I assured them what we were doing would soon work.

After about fifteen minutes, Simmy's batteries finally ran flat. He realized his attentions were not achieving anything and he sloped off into another part of the house. If we were fighting World War II, then I had achieved the D-Day Landings. The most decisive battle had been won, but the war was not quite over. Simmy came back after about ten minutes. He had used his "time out" to evaluate what was going on and decided to check out the situation with another bout of bouncing and barking. This time the bouncing lasted for little more than thirty seconds. The barking lasted a little longer, maybe a minute. Faced with the same noncommittal response, he went away again.

Simmy was going through a routine I have seen many, many times now. He had grasped that there had been a fundamental change in his environment. Each time he was returning, he was doing so in the hope of finding a chink in his aspiring new leaders' armor. I have seen dogs do this a dozen times before giving up. Each time, their energy levels drop a little. By the end, they appear to make nothing more than a pathetic little whimper of protest. The key thing to remember is that it is only when this repertoire is over that you can apply the five-minute rule. Any attempt to get the dog to cooperate with you before then will be ignored.

Soon, Alan and Kathy were replicating my method, using all four elements of Amichien Bonding to establish their leadership over Simmy. In particular, they worked hard at relieving him of the responsibility when visitors came to the house. Here they used different options according to their guests. When an elderly grandmother visited, Simmy was kept in another room. When Alan's brother came, he was briefed to meet Simmy at the door. In every case, however, if Simmy began leaping up and down, they left him to his own devices. At every turn, he was given the same signal: it was not his job to deal with this situation. He should relax and get on with enjoying his life. No one was interested in playing the up-and-down game anymore. Simmy soon got the message, as dogs always do. Soon Alan and Kathy's visitors were greeted with barely a glance. Simmy's bouncing days were over. I am sure he appreciated the rest!

12

Non-Total Recall: Dogs That Run Wild off the Lead

The ability to recall a dog that is off the leash is perhaps the most vital that any dog owner can possess. It can, in some instances, make the difference between life and death. This is one of the key situations when it is vital that the dog sees its owner as a leader who is able to make important decisions and is the most experienced member of the pack.

Over the years, I have seen many instances where a lack of control could have proved fatal. One incident in particular always springs to mind. It happened one morning as I waited outside my doctor's surgery. The building was near a large housing estate and a busy main road. As I waited for the surgery to open, I suddenly saw a Yorkshire terrier racing out of the estate heading straight for the road. The dog was being chased by a group of three children who were shouting and waving at it to no avail. Each time the dog stopped, it looked back at them. As their shouts came closer, he ran off again.

At that time of the morning the road was busy with fast-moving rush-hour traffic. I could see the dog was headed straight for the road. I knew I had to do something so I shouted to the children at the top of my voice. They must have wondered who this lunatic was, shouting and waving at them as if the world was coming to an end. They knew they were in trouble, however, and did precisely what I asked. The first thing I did was to ask them to stand still. I then shouted to them to turn around and run away back in the direction of the estate. To my relief, the terrier saw this and came to a halt, only a few yards away from the road and the busy morning traffic. It then turned on its heel and started chasing back up the road in the direction the children had headed. It was a chilling moment. If they had carried on chasing that dog, I have little doubt it would have been run down in the traffic.

In that case, of course, I didn't have time to explain their mistake to the children. By chasing after their dog, they were participating in its game, and giving the impression that it was leading them. They needed to end that game and reassert some authority. I'm sure that incident was a lesson to them. In reality, getting the dog to understand what is required of it in these circumstances is straightforward. As ever, it requires a combination of persistence and presence of mind to make it happen.

One of the more memorable dogs I have come across was a Saint Bernard called Beau that I was asked to deal with as part of my work on challenging dogs on Yorkshire TV. We all know Saint Bernards are famous the world over for their work in mountain rescues. With their trademark brandy cask around their neck, these remarkable dogs have saved the lives of hundreds of Alpine mountaineers, tracking stranded climbers in the most remote spots and helping return them to safety. It was probably just as well Beau was not working in the mountains, however. He was that rarest of the breed, a Saint Bernard that no one could recover.

Beau's owner, a lady called Heidi, had spent more time than she cared to admit chasing him helplessly around her local parks. No matter what she tried, she simply couldn't recall him to her. She had now got to the point where she had given up on even trying. Whenever she and Beau went out on a walk, he remained tethered to a long lead. She simply could not bring herself to let him run free anymore.

As a responsible dog owner, however, Heidi knew this was far from healthy for Beau. Like all dogs, he needed exercise. I asked her to let him off the lead. He lumbered around the park like a giant tank. When the moment came to recall him, Heidi's efforts were as unsuccessful as she had told me. She called him six times then gave up. Heidi's mistakes were common enough. At her home, I immediately noticed that she had food available for him everywhere. Out on the walk, she was following Beau whenever he was on the loose. In doing this, of course, she was paying homage to his status as leader. She was allowing him to dictate the rules of his own game.

Heidi had firstly to blitz Beau with signals, starting with the four main elements of the bonding regime. It was only by claiming control of her dog at home that she would be able to get him to behave as she wanted outdoors. Beau was basically a good-natured dog, and caught on quickly. It is a tough regime to maintain for many people. But during this period, I ask owners to refrain from taking their dogs out until they are ready. Within two weeks, however, Beau was responding to Heidi's calls to come within her house and garden. She had learned to praise his behavior and he in turn had come to make a positive association with this routine. The crucial thing now was that Heidi reaffirmed the message she had been passing on around the home. She had to assert herself as the person who was going to lead the hunt. It was no easy task.

Beau was getting himself extremely agitated as she got the lead out. So I asked Heidi to calm the situation right down. I got her to place the lead on a table and walk away. The signal was clear: the dog had blown it, the hunt was canceled. Beau had to realize the consequences of his actions. When eventually Beau calmed down, she attached the lead to his collar and led the way through the door. At this stage it was vital that she gained control of the walk immediately. So when Beau began pulling on the lead outside the door, I again asked her to disassociate herself from the situation. She stopped, turned around, and headed back indoors. It took her three or four days before she could get beyond her gate. Beau's persistent pulling meant the walk was constantly postponed. Eventually, however, he got the message and walked on the lead.

The key thing now was to reaffirm the benefits of the recall. I got Heidi to extend her lead even further by adding a long rope to it. I then asked her to begin by extending the lead so that Beau came to a stop six feet away from her. At that point I got her to get him to come to her using food as a reward. Each time he did this correctly, she extended the lead further, two or three feet further each time. Beau responded to her request each time until the rope was extended for a full thirty feet. At that point I asked her to release him.

What I now wanted Heidi to do was to practice what she had been doing on lead, off lead. So I asked her to go through the recall process again. Her hard work back at home was soon paying off. Again, the lure of food drew Beau back to Heidi each time she extended the distance. Soon he was returning to her calls from a distance of more than fifty yards away. Within a month, Heidi's walks with Beau had become the hugely enjoyable experience she had always wanted. Her days of chasing him high and low were over. He would come to her without fail. The outcome could not have been better. And what was more, Beau had become a fitter, healthier, and happier dog altogether.

* * *

If I have learned anything during my time training dogs according to my method, it is that we must always be willing to improvise. The strength of the process I have evolved is its flexibility. It can be modified to suit all lifestyles. As I discovered training a German shepherd of my own, it can also be amended to suit the personalities of all dogs. I have always upheld that the more intelligent a dog the more resistant it is to change. Clever dogs are constantly challenging decisions within themselves. Whatever the activity, they demand to know why they are being asked to participate. It is, I believe, why the brighter dogs fall for my method like a stone. They grasp this is a situation that is of benefit to them and accept it readily.

There are few more intelligent breeds than the German shepherd. And there have been few dogs that have been quicker on the uptake than Daisy May, a German shepherd I bred myself. Daisy May was an irrepressible, hugely energetic dog and a real pleasure to be with. She had proved easy to train according to my method and had adapted perfectly to life within my pack. Then one day, out of the blue, she presented me with her first challenge.

I have always enjoyed taking my dogs out to local beauty spots in my car. One day I took them out to a bridleway in the countryside where they were let loose to enjoy themselves. When the time came to go home, however, Daisy refused point-blank to get back into the vehicle with me. I stood by the car, calling her to me. But all she would do was leap around endlessly, refusing to get in.

Obviously at this point I had the option of simply picking her up and forcing her into the car. As I have explained before, however, I want dogs to make decisions of their own free will. I want them to make positive associations with situations and act accordingly. Simply shoving Daisy May into the car would have been a completely negative association. So I decided to try something else. As she continued playing around, I climbed into the car and drove off without her. In doing this, I was presenting her with a choice. Everything within her told her that her place was with the pack. Her survival depended on it. Was she now willing to live without that pack?

After about twenty yards, I came to a stop, got out, and called her again. Daisy May ran to the car but continued frolicking around. It was clear she wanted to carry on playing this game. Again, she refused to come. Once more, I got back into the car, but this time I drove off at greater speed and over a further distance. My question to her now was do you really want to be on your own? I immediately saw Daisy May in my rear-view mirror. She was barreling along behind me. This time when I stopped, she leapt straight in to join the other dogs. I rewarded her behavior with praise.

My work with other people's dogs has taught me that important lessons like this need underlining as soon as possible. So the next day I attempted to repeat this and went back to the same spot. Once more, Daisy May refused to get into the car at my first request. I was not going to play her game this time, however. As soon as she started playing around, I decided to show her that her actions were going to have consequences. I immediately drove off at speed, heading almost 200 yards into the open countryside. At all times, I should add, we were never less than a quarter of a mile from any road. Once more, Daisy May was in hot pursuit. As I opened the car door, she leapt in. It was the last time I had to go through that particular procedure. After that, Daisy May was always the first dog into the car.

13

Dog v. Dog: Taking the Heat Out of Canine Confrontations

A few years ago, as I tried to find the links between the behavior of domestic dogs and wolf packs, I watched a remarkable film. The documentary traced the story of a community of wolves living in the wild in Yellowstone National Park in Wyoming. Despite the fact that North America is a stronghold of the gray wolf, the species had been absent from Yellowstone's wilderness for more than sixty years. The pack was one that had been placed in the park in an effort to reintroduce the species to the area. The documentary followed their progress as they settled into this environment.

The film was hugely influential in helping me as I put together the ideas that now underpin my method. None of its insights was more useful than those provided in a sequence in which the pack was forced to find itself a new Alpha male. The previous incumbent had been killed, the victim of a human hunter's bullet, leaving only the Alpha female to lead the pack. Soon, however, another wolf from a neighboring pack arrived in an attempt to impose himself. The process that ensued was fascinating. The outsider began by howling to see whether he heard the trademark bass howl of an Alpha male in response. Encouraged by the fact that there was no such howl, he began to prowl the perimeter of the pack's territory.

His attentions soon produced a response from the pack who began going through an elaborate, and at times highly aggressive, ritual. Wolves would take it in turn to charge aggressively toward the interloper. They would pull up short and go through an elaborate routine. It was all posturing. It reminded me of Native Americans throwing a spear into the ground at the feet of a potential foe. Each time, the wolves would retreat before charging again. In addition to all this, there was an immense amount of body language going on.

Throughout this, the outsider remained steadfast. He simply stood his ground, wagging his tail. He did not threaten the other pack wolves in any way. Yet at the same time, he showed no signs of weakness. The pack continued with this repertoire for a staggering six-and-a-half hours. At the end of that time, however, something remarkable happened. Suddenly the charging stopped and the wolves started going to the newcomer one by one. He had faced an all-or-nothing situation. If he had lost, the pack would almost certainly have killed him. But he had triumphed.

Once the pack membership had paid its homage, the Alpha female came over. In one final piece of symbolism, he placed his front leg over her shoulder and his head over her neck. He retained that position for no more than half a second. It was long enough to signal that the deal had been struck. He was the new Alpha male. It was a beautiful sight to behold, an example of nature at its most pure and powerful. The rest of the pack greeted it by leaping around, clearly overjoyed at the fact that order had been restored and the pack had a leader once more.

The dog may have been taken out of the wolf pack, but the instincts of the wolf pack have not been taken out of the dog. Our domestic pets go through their own versions of this behavior on a daily basis. And it is never more obvious than in one of the most common situations dog owners face: when one dog challenges another. Like every other dog owner alive,

I regard the idea of one of my dogs being attacked by another dog as the worst possible nightmare. Dogs are capable of inflicting hideous injuries on each other. It is not beyond them to deliver wounds that prove fatal.

Whenever a dog fights, the physical toll it takes on the animal is invariably matched by the psychological damage it causes the owner. This was certainly true in the case of Christine, a lady I helped as part of my work on television. Christine had begun renting a smallholding in Yorkshire where she had taken in a couple of dogs: Basil, a lively tan-and-white Border collie-type mongrel, and Tess, a small black crossbreed.

The source of Christine's problem was another dog, however. Reggie, a big, tan-colored Rottweiler crossbreed was part of the fixtures and fittings she had inherited when she took over the smallholding. Rottweilers' fearsome reputation is undeserved in my opinion. I have met many lovable examples of the breed. Many people forget they were originally bred as guard dogs by cattle farmers in Germany and Switzerland. Reggie was fulfilling his breed's historic role admirably. Reggie was kept on a chain that was attached to a running pole. Again, this is a principle I cannot condone in any way. Despite the restrictions placed on him, he was more than able to scare off any unwanted visitors. He was a fearsome-looking animal.

Christine's problem was that Basil was one of the few who was not scared one bit by Reggie. On several occasions, he had slipped out of the house, headed for the Rottweiler's part of the yard, and fought with him. We have all come across the Yorkshire terrier that is willing to take on a giant German shepherd, or the dachshund that squares up to the Doberman. While we are all too aware of the size differential, dogs themselves seem to have little or no concept of their relative stature. This is again our human perspective at work. It is we who have diverted dogs along different evolutionary routes. In reality, all breeds are within five re-generations of each other. Given this, it is natural that all dogs regard themselves as the physical equal of each other. Put simply in this case, Basil imagined he too was a Rottweiler. Unfortunately, the advantage in size and power was all too real. Reggie was at least twice Basil's size. Because he was chained, he had also been put in a situation where he had no option but to defend himself. He had inflicted rips, tears, and puncture wounds to Basil's ears, legs, and body; Basil was beginning to resemble a patchwork quilt. Reggie carried a few battle scars too. Slowly but surely, the two were literally ripping each other to pieces.

It is important to say at this point that my method is not going to remove the aggressive tendencies of any dog. As I have explained earlier, the biting instinct is not one that can be unlearned; it is a part of a dog's personality. I sometimes liken dogs to the Rambo character in the *First Blood* movie. Left in peace, Rambo was able to live his life like any normal well-adjusted person. Asked to defend himself, he fell back on knowledge that allowed him to become ultra-violent. Make no mistake, there are dogs that are capable of doing terrible damage to humans in confrontational circumstances. Breeds like pit bulls, for instance, were raised specifically for the purpose of fighting; called upon to do so, they draw on that savage nature to the full. My method cannot remove these basic instincts from any dog, whatever its breed. What it can do, however, is allow people to manage their pets so that the confrontations that bring out this aggressive nature never take place.

Unfortunately, I was not free to help Reggie, as Christine was not able to get his owner's permission so that I could work with him. The owner of the property simply wanted a guard dog on duty twenty-four hours a day. Basil, however, was a different case. The moment I met him, it was obvious Basil was as clear a case of an unelected Alpha as you could wish to see. When I first met him, he displayed all the classic symptoms. He was pulling on the lead, jumping up and barking. He clearly believed he was head of the household. He had even developed the habit of jumping up onto the kitchen worktop so as to be able to keep a lookout through the window.

Christine began by going through the normal bonding process with Basil. During this time, I asked her to be extra vigilant about keeping him away from Reggie's part of the yard. The two dogs were not to see each other. When I felt Basil was ready, we took him out to the yard. I had him attached not only to a lead but to a harness as well. I knew how agitated he was likely to get and did not want to risk the possibility of him slipping his lead. In preparation for what was about to happen, we had put Reggie into a shed.

As soon as we got Basil onto his old foe's patch, however, we let Reggie back out. He remained attached to his chain. At the same time I knelt down, quietly and calmly holding Basil at a distance of twenty feet from the end of the chain. Exactly how Reggie's chain remained in place I don't know to this day. Reggie roared into life and flew at Basil. Basil was as ready as ever for the confrontation, however; it was all I could do to hold onto my dog. The two dogs were ready to tear each other to pieces. But as long as they were signaling aggression toward each other, I made sure they would not physically be able to reach each other.

Eventually, the adrenaline levels subsided and tiredness set in. It was not the six-and-a-half hour ritual wolves go through, more like a quarter of an hour. The moment the threatening behavior stopped, Christine, as we had arranged in advance, appeared with a bowl of food for each of the dogs. The signal we were sending out was two-fold. We wanted the dogs to make a positive association with each other's presence. And we wanted them to understand that positive association only occurred when they were being peaceful.

I cannot report complete success in this case as yet. The two had been fighting for a long time. This was no quick-fix situation. Basil responded well to the bonding process and began to be much calmer in his confrontations with Reggie. The two have not had a fight for some time. Basil has not needed stitches for some time. If the Rottweiler too had been given the right signals, I have no doubt the two would be able to peacefully co-exist. This has not yet happened, however. The best I can hope for is that Basil remains a stranger to his local vet for years to come.

* * *

Whenever we jump in our car, we face the reality that, no matter how expert a driver we may be, there is a risk we will be confronted with another motorist who is unfit to be on the road. The same reality confronts every dog owner every time he or she steps out of the safety of their home. In the main, walking a dog is an enjoyable—and at times highly sociable—activity. I have forged many a friendship while out walking my pack. Yet it is a sad fact of life that most dog owners will come across a situation where their pet is faced with the aggression of another animal at some time.

There is, it is sad to say, nothing we can do about this. Not all owners exercise the care and control I encounter in most of the people I work with. Every decent dog owner is the victim of an irresponsible one at some time; we have to recognize that. Apart from anything else, as I have said before, we cannot remove the natural defensive instincts a dog falls back on when it cannot escape a confrontational situation. My best advice is that we should all avoid and ignore any such situation to the best of our ability.

There is much that can be done to ensure our dogs are not the aggressors, however. Again, the central ideas here are rooted in the nature of the animal and in the dynamics of the wolf pack. In the wild, wolf packs do all they possibly can to avoid other packs. The intensity with which they mark dens and hunting trails is designed to ensure packs remain confined to their own territories. Confrontations are rare.

When we bear this in mind, we realize how unnatural it is for domestic dogs to come into contact with other packs. We must remember here again that, to a dog, a pack can consist of as few as two members: a human and another dog. To a dog that believes it is the leader of its pack, it is a moment of potential danger. If a confrontation occurs, it will do anything necessary to protect its charges. The anxiety may be heightened if such meetings occur in a dog's familiar walking environment, its local park for instance. On top of its responsibility to its pack, a dog may sense some territorial threat too.

To overcome the natural anxiety that occurs, I recommend all dogs I deal with go through a process I have called "cross-packing." It is something that can be developed as owners learn to take charge of the walk. The idea is to get the dog used to coming into contact with other dogs and their owners, so that their packs cross without incident. The long-term aim is to make dogs as impervious to others as modern urban man is to his fellow humans. Whenever a dog comes into contact with another, I ask owners to simply ignore the other dog. If the dog responds to their example by letting the other dog pass without any reaction, it is rewarded with food. Again, the dog is being encouraged to make a positive association with this situation. The key to ensuring this is a painless process lies in the principles that will already have been learned in the home. Most importantly of all, owners must demonstrate leadership qualities that the dog can identify and believe in.

As I have said, however, no matter how much control an individual owner is exercising over his or her dog, there is nothing he or she can do to control the behavior of other people's dogs. I am often asked what clues people should look for in the body language of aggressive dogs. People, understandably, want to know how to deal with the inevitable situations where one dog challenges another. What makes a growling dog turn into a fighting dog, what are the triggers to the attack, and so on? My answer is always the same: they should look to the owner rather than the dog—leave the dog to weigh up its own kind.

If an owner is relaxed and happy, his or her dog invariably feels the same way. On the other hand, if an owner is waving his or her arms around, looking stressed or aggressive, and struggling to hold onto their dog, then the chances are the dog, too, is in a highly combustible state. A dog accompanied by this type of owner is far more likely to attack. An attack by another dog presents perhaps the greatest test of an owner's leadership. My advice is to avoid confrontations at all costs. I certainly advise against aggravating the situation by berating the other owner. It is imperative that someone remains calm. Again, it is a time to think of Kipling, and to keep your head.

I am often asked why I do not recommend people simply pick their dog up in such situations. My reasoning here is that it sends out confusing signals to the dog. Firstly, it is being removed from the level of its fellow animal and cannot, therefore, assess the situation for itself. Secondly, the owner risks getting bitten in the process. Far better in my opinion is to show strong leadership and guide the dog to a way of dealing with the situation which it will be able to make again and again if necessary.

There is no doubt that anxiety about potential aggression between dogs can ruin the life an owner enjoys with his or her dog. The case of a retired nurse, Miss Artley, exemplifies this better than any other I have come across. Miss Artley lived in a lovely cottage in the seaside resort of Bridlington. She shared her life there with two beautiful Old English sheepdogs called Ben and Danny. Unfortunately, the dogs had become very aggressive toward other dogs during her daily walk. At well over one hundred pounds in weight, the dogs were big. In comparison, the diminutive Miss Artley weighed no more than ninety-eight pounds. She could barely control Ben and Danny at the end of their leads, and picking them up was certainly not an available option. As a result, she had become increasingly helpless to stop the attacks.

By the time the owner called me, matters had deteriorated so much that the poor lady had resorted to walking the dogs at the most unseemly hours of the day. She told me she was going out at midnight and then again at five in the morning, so as to avoid anymore upsetting confrontations. I am sure, like many people, she had her reservations about my abilities before I met her. I can understand that perfectly. Fortunately, I converted her within five minutes.

Miss Artley kept her dogs in the garden because they were too boisterous. Her home was impeccable and their exuberant charging around was forever sending her prized possessions flying. Within five minutes of arriving, I had succeeded in calming them down. As ever, I had entered the house starting as I meant to go on, sending out strong signals that I was the leader and my authority was absolute. The two dogs were soon lying contentedly in the living room for the first time in their six years with their owner.

The owner's major problem, however, was the walk. My solution here was simple. My goal was to ensure she would avoid situations where the dogs would confront other animals. To this end, I asked her to use food rewards to get the dogs used to moving away whenever they came into contact with other dogs. So, for instance, if the owner was walking down the street and saw another dog approaching, I asked her to avoid a face-off by crossing the road. Once safely across, I asked her to reward the two dogs with a tidbit. This simple action not only removed the negative possibility of a confrontation, but also showed the dogs that the owner had taken the decision to lead in the defense of the pack. At the same time, I stressed to Miss Artley the importance of her remaining calm throughout this type of situation.

This is not a problem that is solved in an instant. And the importance of achieving the bonding process before attempting a walk is paramount. I have in severe cases asked for a dog to be confined to its home for a week before it goes out into a potentially confrontational situation. These confrontations occur because dogs believe they are repelling a potential attack on a pack for whom they are responsible. If they have been demoted within the order of that pack, they will defer more easily to their new leader.

Miss Artley stuck rigidly to what I asked of her. Within two weeks, she was out walking during normal daytime hours. The transformation in her life was obvious when she rang me a year later on the anniversary of my visit to her. She told me that she, Ben, and Danny had just returned from the beach where they had been walking and playing with some canine friends of theirs. Their re-integration into Bridlington society was complete.

14

Tales of the Unexpected: Fear of Noises

People often question what is wrong with a dog believing it is the leader because, as humans, we are taught that self-esteem is an important, empowering force. By relieving a dog of its rank, they ask, aren't we stripping it of its self-esteem, its self-confidence? If the world we inhabited had been created by dogs for dogs, then the answer might be different. The fact is, however, that dogs live in a world exclusively geared to the needs of humans. That is where the problems begin. And that is why the answer to that question must be a resounding "no." The dog's belief in the hierarchical system from which it has evolved is absolute. If it believes it is leader, then it is also convinced it knows more than any of its subordinates. Its logic is simple. If a junior member of the pack knew more than it, it would be the leader! As long as a dog believes it has this status, it will take on the mantle of decision making in every situation it faces. The reality is that it is extremely dangerous to allow a dog to do so; in an unfamiliar situation, a dog will make up its own rules as it goes along.

The obvious comparison again is with young children. No matter how smart the child, no matter how self-confident he or she may be, would a parent allow a five-year-old to drive the family car or to take charge of a shopping expedition in the middle of a city center? Of course not; a child is simply unable to deal with the situation. The difference is, of course, that child will one day grow up. Dogs, as I have explained, remain puppies for life; they can never be given this responsibility.

The danger of allowing a dog to believe in its leadership status is never more acute than in situations in which it is confronted by sights and sounds it does not understand. It perceives these situations as potential dangers to its pack members. As anyone who has seen a dog chase after a car or become distraught at the sound of thunder will know only too well, the reality is that they are a far greater threat to the dog.

I have been asked to look at many cases of this kind. They have ranged from dogs who go berserk at the sound of a passing car or truck, to pets who howl and bark continuously at the sound of thunder and lightning or the exploding of a firework. Any of these can cause a dog an enormous amount of distress. I am sure we have all heard stories of dogs who have run into an open road at the sound of a car backfiring, and have been run over. It is a hugely important problem area. In each case, it is a symptom of the same problem: the dog's inability to cope with leadership. What makes this situation more dangerous than most is the fact that as well as being unequipped for its responsibility, the dog senses it is out of its depth. Its reaction is, quite simply, to panic.

Much of the knowledge I now have came from working with my own dogs. A few years ago I used to dread November 5th, Bonfire Night in the United Kingdom, and the noisiest evening of the year. Over the years, my home, adjacent to a showground where the local council organizes the area's main fireworks display, had become something of a refuge for traumatized dogs. As the fireworks exploded one night a few years back, I recall I was summoned by a frantic knocking on the door. A passer-by had found a dog sitting in the middle of the road outside my house, literally paralyzed with fear. The caller had assumed wrongly it was mine. There was no sign of an owner. I had to smile when I saw a man trying to coax the dog with a biscuit. No food on earth was capable of taking this poor dog's mind off the terrible noise erupting all around it. I carefully picked the dog up off the road and brought it into the house. I later discovered her name was Sophie. She sat petrified in my kitchen for hours. I just left her alone, offering her food and drink. It was three days before her owner reclaimed her.

It was much the same the following year, when a black-and-white Border collie was brought round. She had clearly run away from her owner amid the explosions. I calmed her nerves by putting her in my car with the engine running and the radio blaring until the display had finished. Fortunately, her owner found out where she had been taken and was relieved when he collected the dog later that evening.

The dramas were not confined to other people's dogs, however. The display also used to terrify my little beagle, Kim. I think the first time it happened, I simply sat there cuddling this poor pathetic trembling creature. Another year, I loaded her and the rest of my dogs into my car and drove into the heart of the Lincolnshire countryside to be away from the explosions. My reaction was, I now realize, precisely the same as it had been when my children had woken up in the night, frightened by the sound of thunder and lightning. At times like this, our natural instinct is to gather our loved ones around us and to comfort them. Instinctively we are all acting out the scene from *The Sound of Music* when Julie Andrews gathers the young Von Trapp children around her and starts singing about her favorite things. I remember I used to tell my children that thunder was the sound of the angels playing skittles!

It was only as my method began to evolve that I realized the awful mistake I was making by replicating this with my dogs. What I was doing was praising my dogs' behavior in reacting to noise. What I needed to do was the complete opposite, to ignore the whole situation, and show it was nothing of consequence. It all fell into place as I understood the dog's absolute belief in leadership. If a dog has elected its owner leader, it will always implicitly believe that leader knows more than it does; if he or she didn't, they wouldn't be leader. I realized that what I needed to do in instances like this was to display disdain for the whole situation, to remain calm and simply ignore the noise. It was back to Kipling again; the leader must keep his or her head while others lose theirs. I realized that if a dog believed in its owner and they were ignoring the noise, the dog would do the same thing.

The principle really proved itself when I worked with a similar problem, that of a dog that was frightened by the noise of cars. In my experience, the sound of a car or truck engine roaring within a few feet of its face can be one of the most frightening and disconcerting things a dog has to deal with. I have met owners who are unable to take their dog anywhere near situations in which it will encounter traffic of any kind. For those who live in built-up areas, it can condemn both dog and owner to a kind of imprisonment.

Soon after I had begun using my method, I was approached by an elderly gentleman who was having severe problems walking Minty, a very pretty blue merle Border collie he had taken in on behalf of his son who was working overseas. Every lunchtime and evening, this gentleman liked to visit his wife, who was resident in a nursing home nearby. The problem was that his visits to her were being disrupted by the fact that Minty panicked completely whenever she saw or heard a motor car. The walk to the home took them along an extremely busy road. The owner had been forced to turn home on more than one occasion, and was becoming increasingly distraught about the situation.

I set to work with the owner at his home, going through the four elements of bonding first. It is worth mentioning at this point that the work I do is always done in the home environment if at all possible. There are two reasons for this. Firstly, the dog is far more likely to be its real self at home; once you take a dog off its patch, it will behave completely differently in my experience. Even the most contented and confident dogs can become terrified when they go outside. The other beauty of working at home is that the owner sees everything you do. There is no suggestion of secrecy or mystery to what I do. Invariably too an owner feels more relaxed at home. And the more relaxed he or she is, the more likely it is that the process will go ahead smoothly.

In this case, the owner grasped the main areas of the bonding process well. But it was clear that the key area with Minty was going to be the walk. The strategy I devised was based on a simple idea. When Minty went out on the road, I wanted it to be an experience with which she made a positive association rather than a negative one. Therefore, after an hour or so building a relationship with Minty, and with my status as leader established, I attached the lead and took her off for a walk.

The road was busy with traffic, precisely as I wanted it to be. As soon as the dog started reacting to the first passing car, I said, "Minty, come," offering her a small piece of cheese as I did so. I did the same thing with each subsequent car. If Minty failed to come to me and continued barking at the car, I ignored her. I was not going to treat undesirable behavior positively. But if Minty came to me, I rewarded her with cheese and gentle praise. I continued doing this as we carried on walking down the road. It was a busy street, and we hadn't got very far at all before Minty was looking at me rather than the road whenever she heard the sound of an approaching vehicle. By the time we had been passed by a dozen cars, I was able to dispense with the food association. We had been out for only a quarter of an hour. It was simple. I had made a good association out of a bad one. I handed Minty back to the gentleman, and he was soon making his way to the nursing home to share the good news.

* * *

Of course, it does not require a car backfiring to drive a dog to distraction. In cases like Bonnie's, a black-and-tan cross corgi-Border collie who lived with his family in Revesby, Lincolnshire, even the ringing of a telephone could be the cause of enormous anxiety. As is so often the case, Bonnie's owner Pat called me for a variety of reasons. Bonnie was suffering from many of the symptoms of nervous aggression: pulling on the lead, jumping up, and barking. It was while talking to Pat that I learned of something very specific to Bonnie: her reaction to the phone. Pat told me that whenever she heard it ring, Bonnie would become frenzied, panting, rushing up and down, and even crying. Her reaction had become so extreme, she had even begun going through a strange ritual where she licked the carpet until the ringing stopped—and for a full fifteen minutes afterward!

I was interested to see this for myself and decided to test Bonnie's reaction by visiting Pat's house and calling her from my mobile phone while I was in the same room. Sure enough, Bonnie went into a complete spin. The exercise helped me learn as much about Pat as Bonnie, however. I saw that Pat was chiding her dog, and would shout "stop it" at the top of her voice. As I talked to her, I was not surprised to learn that Pat had also got into the habit of rushing to the phone whenever it rang. Of course, all these things were simply exacerbating the problem.

Bonnie's anxiety was rooted in her belief that she was the leader of her household "pack," and the sound of the telephone ringing represented an unknown threat to it. Bonnie's inability to negate or deal with that threat was sending her into a blind panic. Pat's highly excitable reaction was only adding to the tension. Bonnie's licking of the carpet was her own highly obsessive demonstration of her hopelessness. My task was firstly to take all the drama out of the situation, to begin to persuade Bonnie that the sound of the phone was nothing worth worrying about at all.

From the moment I arrived, I gave Bonnie signals consistent with my leadership. Satisfied that she saw me in this light, I put her on a lead, sat down calmly with her, and then began calling Pat's number again from my mobile phone. When the ringing began, I remained totally relaxed. I did not react in any way for several rings. Bonnie was anxious but soon realized there was something different happening. To encourage her to stay calm, I rewarded Bonnie with a special tidbit, a piece of cheese. The idea here was to desensitize her, to help her to make a positive association rather than a negative one whenever she heard the familiar ringing sound in the future.

Bonnie reacted well and, while agitated, remained at my side, under control. During the space of the next hour, I tried the same thing again every fifteen minutes or so. When the phone rang for the fourth time, Bonnie did not react at all. The frenetic behavior of earlier had disappeared, as had her habit of licking at the carpet. Her attitude to the phone has remained the same ever since.

It took three of my own puppies to finally drive home to me the particular message of good associations. My German shepherd, Sadie, Sasha's daughter, was approaching one year old while Molly, a little springer spaniel, and her half-brother, Spike Milligan, were seven and five months old, respectively. As they approached their first Bonfire Night, I had made all the preparations I could to make sure they were not distressed. I had kept them indoors and had put a little television set on in the kitchen where they were eating and resting. The idea was that the noise from the television would act as a useful distraction when the fireworks began.

I was so wrapped up in other things, however, that I forgot to close the door when I went out into the garden to watch the fireworks. Before I knew it, all three puppies were bounding after me. Their timing could not have been worse (or better!). Almost immediately the first rocket of the evening screeched up into the dark sky above us and exploded in a blaze of color.

I didn't have time to admire it; as soon as the explosion happened, Spike in particular panicked. He dived to the ground and wrapped himself up around my feet. At the same time, the other two stood there, crouching close to the floor, looking wide-eyed at me for guidance. By now I was experienced enough to know I had to act decisively. So I simply smiled and said, "That was a big one wasn't it," in a matter-of-fact tone of voice, and carried on doing what I was doing. It was enough for the dogs to relax. Moments later, all three of them had picked themselves off the floor and began to move away from me. They spent the next half hour watching the remainder of the display. The following year when the fireworks started, they were scrabbling at the door trying to get out. I think it is their favorite night of the year now.

15

New Dogs, Old Tricks: Introducing Puppies to the Home

Much of the work I do is with remedial dogs, animals with behavior problems ranging from pulling on the lead to destroying the home. Invariably the root of these dogs' problems lie in the past. Their owners, through no fault of their own, have spent years giving them signals that have in turn given the dog a misplaced sense of its own importance. My task is to redress this balance, to offer signals that establish a new order and a kinder, calmer future for both the dog and its owner.

It does not take a genius to work out that the ideal way to avoid such problems is to deal with a dog at the very beginning of its life. A puppy offers the perfect opportunity to start as we mean to go on. Some people are surprised to learn I am often called out to help owners with their new puppies. The truth is, I really welcome these cases. By definition they are coming from the ideal dog owners, people who care, who respect and want to understand their pet from the beginning of its life with them. To live with animals, people should learn about them beforehand. Far too few people take the time and trouble to do so.

I must say that I have strong views on who should and should not be given puppies. Quite simply, many people are unfit to look after dogs of any kind let alone vulnerable young animals. Puppies should certainly never be given to young children as a present. I make no apologies for saying this. If a child wants a plaything, then I suggest their parents give them a doll or a train set. A dog is not a toy.

My opinions on this have upset people in the past, I have to admit. It is very rare indeed that I agree to give people a puppy when they first visit me. I prefer to be certain that a dog is going to the right sort of home, and I have to be firm about that. I recall refusing to give a puppy to a family who had driven 200 miles to see me once. On another occasion, I refused to part with a puppy that a family wanted for Christmas. They wanted it for their children and, when I refused, their initial reaction was that they'd go elsewhere. Of course, they would have found someone who would have sold them a puppy. There are people who will breed or sell dogs without any concern for the animal's welfare. In this case, however, they understood my motives. My argument against giving dogs at Christmas is simple: calmness and consistency are central to everything I do. Christmas is the least calm, least consistent time of the year.

The family talked about it. I am glad to say they understood what I was saying and agreed. Rather than having a present on Christmas Day, the family came to my house on Christmas Eve instead. The children got the excitement of seeing their new friend but understood they would have to wait until after the holidays, when everything returned to normality, before they could return to take it home with them. Apart from anything else, this would ensure that they were genuine in their desire to take the dog and that they would train the puppy in the right environment. I handed over the puppy in the New Year, happy it was going to a good home.

* * *

There are one or two golden rules about taking in a puppy. The first is that the dog should be no younger than eight weeks when it leaves its original home. My reasoning here is again connected to the dog's nature. All puppies are born into a natural family environment, the litter. It is here that it must learn the fundamental facts of life. It has to develop social skills within the litter and it has to learn the language of its peer group. To take it away from the litter environment before these first intense eight weeks are over is, I believe, hugely damaging to a dog.

Once the puppy has left the litter, it is the first forty-eight hours in the new home that become the most crucial. It is a harsh but important truth to remember, but the fact is a puppy is a pack animal that has been removed from its pack. The litter should be a happy, lively, and loving environment where it is interacting with its siblings. The dog is being transported into a completely alien environment, a new home it has had no choice in selecting. Treating the puppy as you would any normal dog is potentially traumatic. It is going to be a nerve-racking experience for the puppy, no matter how loving a home it goes to. For this reason, I believe in establishing the closest possible bond with the puppy during these two days.

I believe in doing all I can to ensure they like their home environment and to make life within it seem as natural as possible. To this end, I actually advocate sleeping with the puppy on the first night. I am not saying it should come into its owner's bed. A far more practical method is for him or her to lie alongside the puppy on a covered sofa for the night. It is a small sacrifice to make as, in my experience, it reassures the puppy at a particularly vulnerable time. The bond this establishes will also help the next day as you help it in investigating and exploring its new environment. It is vitally important that the dog feels comfortable here. This is where it is going to get its food, this is where it gets its affection, this is where it is going to bed down.

At the same time, however, it is important to establish good habits immediately. For reasons I will come to, I do not find gesture eating is necessary with puppies. The remaining three elements of Amichien Bonding should be introduced as early as possible, however.

The most important element of all is undoubtedly establishing order at times of separation. Tempting though it is when owners come in from shopping and this lovely bundle of fluff comes bounding up to them, it is imperative that owners ignore the puppy in these early days. The signal being sent out must be clear and unequivocal: "I will play with you but not now. I will let you know when." It must be sent out from the very beginning, from the first separation even if it has gone into another room and has been out of an owner's sight for a few seconds.

The two processes may seem contradictory. How can an owner be authoritative and loving at the same time as he or she is enforcing such strict rules, people often ask. I point out that the joy that comes when the puppy learns to play on the right terms is, if anything, even greater than that one would get in a less regimented household. There is no question of the fun being eliminated; quite the opposite. It is simply that the affection must be given in the right direction.

The good news with puppies is that the five-minute rule I suggest owners use after separation is almost always enough time in this case. In grown-up dogs with behavioral problems, the repertoire of tricks they will use in trying to get attention can last any length of time. I have seen it last from ten seconds to an hour and a half. An adult dog can leap around, bark, and whine for a seeming eternity. A puppy doesn't get to that stage.

Once the puppy has settled down, the normal process of getting it to come to its new leader can begin. And, as I say, it is here that the true enjoyment can come in. Part of the fun of getting a dog is choosing a name. It is vital that this name is used from the very beginning. At this stage, the more familiar owners are with their dogs the better. I ask owners to call their puppies to them as often as they can, always remembering to reward them with tidbits and praise when they do the right thing. As far as I am concerned, there is no limit to the number of times a puppy can hear the words "good dog."

One of the great joys of training a puppy is the speed with which young dogs learn new tricks. I have found that if you repeat any procedure three times, a puppy will pick up the message, whatever it may be. As with older dogs, it will be clear to see when the Amichien Bonding is working. When it begins standing wagging its tail or sitting in a relaxed position while it waits for your attention, the puppy has confirmed the leadership election process is working. As this develops, owners can also begin working on the other areas of bonding. I do not recommend taking puppies out for walks until two weeks after their first sets of vaccinations, that is until they are about fourteen weeks old. Puppies are simply unprepared for the big wide world at this point. It is far better in my experience to put them into a well-run puppy playgroup, where they can run around in a situation similar to the natural playfulness of the litter environment.

At the same time, however, it is important that the principles of heel work are established early on, and I recommend training the puppy in the home or the garden. The important thing is to teach the puppy that the best place to be is by its owner's side. Again, it must be done through food and reward. If the dog wants to walk ahead, the lead must be relaxed and the dog must be returned to where it should be. Tugging matches should be avoided at all costs. There is nothing a young pup loves more than a game. There will be more than enough time for games later. For now, it must learn the rules of a different game. If you don't lay down those rules at this point, believe me, it will make up its own.

To my mind, the tone of voice an owner adopts with a new dog is of paramount importance. I ask people not to shout or shriek but to make what I call a bonny sound. I remind them that the dog is supposed to be man's best friend. How do they talk to their best friend, do they shout and bawl or do they talk kindly and calmly to them? Once the dog is responding to gentle commands, the voice can be reduced to a near whisper. This will really bear fruit later on. A dog that is tuned in to soft commands will really pay attention when the owners raise their voice.

In the case of decision making at the door, the puppy should be disregarded when people come in. It can work two ways in this situation: in some ways, it is easier to ignore a small dog, and on the other hand, there is nothing more certain to arouse visitors' sentimental streaks than the sight of a cute puppy. It is imperative that the principles are adhered to at all times, however. How often have we all heard that saying that a puppy is not for Christmas, it is for life? Well the same applies to my method. It is not something to be picked up and discarded. Owners must start as they mean to go on, then stick to it.

I have talked about the power of food already. It is nowhere more useful than in puppy training. In this case, however, feeding methods have to be subtly amended to take account of the unique circumstances at work in puppies. The central message of feeding remains, as ever, leadership. An eight-week-old puppy is generally on four feeds a day. In bringing its food to it this frequently, owners are also delivering a powerful and consistent message. They are the providers, the authority within this pack lies with them. Given this, I see little need to carry out the normal gesture-eating technique as well. Why use a sledgehammer to crack a nut?

At the same time, however, food plays a really useful role in teaching other behavior. One of the simplest is teaching the dog to sit. As I have said before, teaching a dog to adopt a sitting position is a priceless tool to have available. By using the method outlined earlier, and bringing food up to and then over a puppy's head, the dog will quickly learn to do this. Once more we are playing on the "What's in it for me?" principle, the self-interest that is already ingrained in the puppy. It never ceases to amaze me how quickly puppies cotton on to this.

16

Gremlins: Dealing with Problem Puppies

Puppies can provide the perfect opportunity to train a dog correctly from the very beginning. Sadly, not knowing the right way to introduce a puppy into a home can bring disastrous results. I am often asked to deal with puppies that have become unmanageable, and arrive at a home to discover a scene straight from the film *Gremlins*. Weeks earlier, the owners were cooing over their adorable new, fluffy friend. By the time I turn up, they are living in fear of a creature that—as far as they are concerned—has suddenly turned into a monster. The truth of the matter is that it is as easy to create a badly behaved puppy as it is to train a good one.

When people ask me how they can train a dog so it is happy and well balanced, I often begin by asking them to turn the situation on its head. If they wanted to deliberately go out and create a completely screwed-up young dog, how would they go about it? They'd probably talk to it in a language it didn't understand, ask it to do a job it was not equipped to do, and spend their days offering contradictory signals that ensured it didn't have a clue what was right or wrong. One moment they'd reward it for being an exuberant ball of fun, the next they'd chastise it for the same behavior. That is precisely what many owners do with their puppies. What they need to do is the complete opposite of these things. The reality is that any fool can wind a dog up; it takes a genuine dog lover to create a happy and contented pet. Two cases exemplify the main problem areas I am asked to tackle in the case of puppies: teething troubles and toilet training. Both are caused by owners heading off down the wrong path at the beginning of a dog's life.

Of all the problems people experience with puppies, by far the most common is teething troubles. Once more, it is useful to understand a little of the natural forces at work before exploring this subject. Puppies develop an armory of little needle teeth at an early age. They have no real function apart from allowing the dog to test the power of its jaws. Puppies, much like little children when they first acquire teeth, do this by biting everything and anything they can place in their young mouths. Within the litter environment, they bite their brothers and sisters. The siblings deal with this with a simple signal: they squeal then walk away from the situation. In the absence of siblings, however, a puppy being raised in the domestic environment will happily bite whatever it can slip into its mouth, including its owner's fingers.

To my mind, the best way of dealing with this problem is via play. There is no place for pain in the training method I undertake. Far from it, I much prefer to teach dogs the important lessons of their young lives through fun and games. Puppies present an ideal chance to do this, provided it is done in the right way. I always advocate that puppy owners have a plentiful supply of toys and objects that it can chew on. They are the equivalent of a baby's teething ring. Puppies are teething for fourteen months, so they have got to be helped. The choice of toys is entirely up to the owner; they can include things like chew sticks and rope-like raggas, or even a knotted towel dampened down. My only request here is that the toys are of reasonable size as small objects can easily slip down a puppy's, or even a grown dog's, throat.

My dad, with his dog Gyp.

Making friends—me,
at the age of four,
at a family picnic in
Norman Park, Fulham.

First love: me and Shane, the Border collie that inspired my passion for dogs.

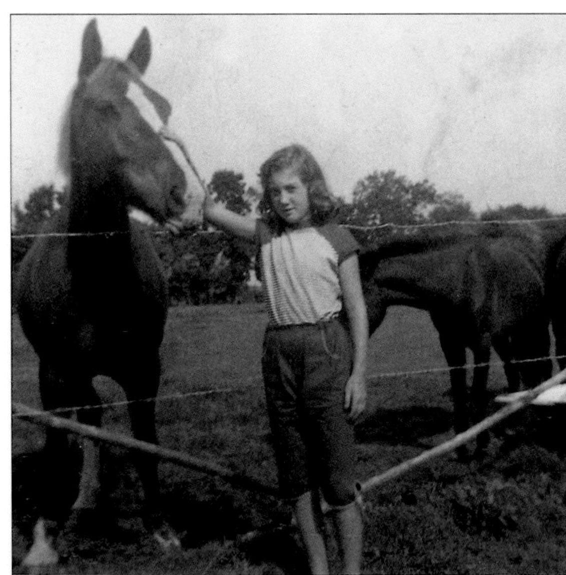

Horsing around at the age of ten.

Show-stopper—me with my prize-winning springer spaniel, Khan.

Family. My son Tony (right) and daughter Ellie, with our gun dog, Kelpie.

(Above) Dan Broughton riding Ginger Rogers, the horse Monty Roberts transformed in twenty-three memorable minutes.

(Right) Donna, "The Duchess."

Barmie by name… me with the rescued Jack Russell that taught me so much.

A young Sasha, my black shepherd, tries and fails to get Sandy's attention.

Sasha teaching Barmie how to play a game of tug.

An Alpha wolf shows its superiority by arching its neck over the vulnerable head and shoulders of another.

Just as wolves circle their prey, so my spaniels demonstrate exactly the same instinct.

The similarity between wolves and dogs can be astounding. Here, a wolf pup asks for food while, below, Molly's puppy does the same.

Both wolves and dogs play at pinning down their prey.

These toys prove invaluable when a puppy starts to chew on an inappropriate object, say the tassels on your furniture. At this point, I recommend the puppy is distracted by one of its toys which is then thrown somewhere else for it to play. The important thing here is that the puppy's natural exuberance is not being punished. The owner is redirecting play in a positive way. If the dog behaves, the game is brought to a close via the "thank you game": the toy is taken away from the puppy, the dog is rewarded and told "thank you." It is another simple way of underlining the message of Amichien Bonding. As leader, the owner has selected the toy as well as when the game happens, how long it continues, and when it ends.

Obviously if a puppy oversteps the mark, then leadership has to be asserted. Puppies are, for instance, very fond of pulling at items of clothing and biting. This is something that must be nipped in the bud. What I do to teach a puppy to pull its punches is, if a puppy puts its teeth, however lightly, around my arm, I yell out and walk away to disencourage harder biting. If a puppy persists in misbehaving, I ask that it is placed in isolation, ejected from the pack for about five minutes, so that it has time to calm down before being silently readmitted to the group.

It is all too easy for owners to send out the wrong signals to teething puppies, however. Such was the story in the case of an Akita puppy called Nuke. When I went to see Nuke's owners, a mother and her three children, they explained that Nuke loved to play a biting game. The whole family would put toys or their hands up to his mouth and let him nibble away. If he nipped at them, they would tap him on the nose. It had all seemed great fun at first. Unfortunately, Nuke had become more and more exuberant in his playing of the game and had begun hurting the children. He was biting more and more ferociously each time.

Akitas are majestic, beautiful but powerful dogs even at that age. He had drawn blood from all the children. Nuke was only eleven weeks old. The family had already begun locking him away in a separate room in the house. Talking to the family, it was clear they had made a number of mistakes. In particular, by indulging Nuke's natural desire to exercise his teeth, the family had made a rod for their own backs. The puppy had learned how to make the owners give him attention on demand. He had also begun to learn how to manipulate them, at playtime in particular.

As I have explained, it is vital that the leader has control of the playtime. The leader must decide what the game is, when it starts, what the rules are, and when it ends. Nuke was making all these decisions. This had to be changed. My first task was to set about re-establishing the leadership. The children were all teenagers and able to comprehend my method's principles, but because the house was very busy, with other children popping in all the time, I asked them to keep Nuke confined to a specific area when they were not alone.

They kept Nuke behind a kitchen gate. When the family was alone together, they allowed him to come back into the living room. Each time he would come bounding in, but each time they would block him with their bodies. If he jumped up in the old way, expecting them to indulge him in his biting game, they just took their arm away. If he did succeed in biting them, I asked them to yell and walk away, precisely as puppies' siblings do in the litter. Nuke quickly cottoned on to the fact that he was not getting the attention he wanted. A dog is no different to a human in this respect; if something isn't getting the desired result, it stops doing it.

There was no uninvited activity; Nuke quickly learned that he had to be even-tempered, well behaved, and had to exercise self-control. And as I have said before, the most powerful form of control is self-control. Within a few weeks, Nuke's behavior had improved enormously. The children were able to play with him much as they had before. The difference this time, however, was that the rules of the game had been changed.

They decided when, where, and how long the activity lasted. Nuke was back on the path toward being a well-balanced dog.

* * *

The second most common problem I am asked to tackle with puppies is their toilet training; it can become a very stressful event for both owners and their dogs. In the summer of 1997, I was asked to visit a family having problems with D'Arcy, their black-and-tan Gordon setter puppy. D'Arcy was every bit as aristocratic as his name suggested. Even at the tender age of five months, he was a beautiful, noble-looking dog. He was clearly going to be a magnificent adult. To his owners' acute embarrassment, however, D'Arcy had started eating his own feces. The family had tried all they could to rid him of the habit, but the harder they tried, the harder D'Arcy worked to evade detection. By now, he was hiding away in corners of the garden and slipping under bushes to do his business. The family were deeply distressed about it and had no idea how to tackle the situation.

No sooner had I met D'Arcy than it was obvious he had several recognizable problems. As young as he was, D'Arcy was clearly stressed. He was jumping up and pulling on the lead, he was constantly "in your face." To the family, these were not even symptoms, but to me, they were all related to the central problem. He was already convinced he was the leader of this particular pack. As I spoke to the family at length, it also became clear why toilet time had become the real focus of his anxiety. The family were fastidious, house-proud people and had become almost neurotic about his going to the toilet. If they thought he was going to go, they would pick him up and rush him outside, making an enormous fuss as they did so. If droppings were discovered in the house, they would go through a similar theatrical scene.

It was clear to me that D'Arcy was stressed because not only did he believe he was leader of his pack, he also felt he was failing in that role. Part of his job was to keep the family happy. He was clearly failing in this so he had addressed the cause of the unhappiness, the product of the traumatic toilet time, by eating it. My task was two-fold. In addition to removing the leadership from D'Arcy, I also had to take the drama out of toilet time.

Toilet training is, of course, a fundamental part of puppyhood, one that has spawned a host of conflicting ideas. Some of the traditional methods such as rubbing the dog's nose in its feces border on the barbaric. It has no place in my method. Yet there is no escaping the fact that it is a practice that must be learned. In my experience, there is no need to do it by giving a puppy a lecture on etiquette.

I began instead by getting D'Arcy's family to start the bonding process in the normal way, ignoring the puppy's attentions. He was a demanding dog, so this took some time but produced good results. To improve the situation at toilet time, I got them to encourage D'Arcy's behavior through stimulus and response. They were obviously on tenterhooks about him going to the toilet. I explained that it had to be a hit-and-miss thing. They were never going to catch him every time. I asked them to concentrate on the most likely toilet times, first thing in the morning, when waking up after a sleep, and after meals. The most important thing, however, was that they calmed the whole process down, that the drama was removed from it. Rather than running around in a fluster, I asked them to be relaxed and happy. And, as ever, I wanted them to be consistent in what they did so that D'Arcy understood what was in his best interest.

The first job was to stop him eating his feces. So whenever one of them was at hand as he went to the toilet, I asked them to leave him to finish, then get him to come to them with reward. I got them to come up with a consistent piece of praise, telling him "clean dog" as they stroked him and gave him his reward. As D'Arcy digested his reward, they would be free to dispose of his droppings.

It is worth mentioning at this point that toilet training is one of the rare occasions when the owner can go to the dog with a reward. In my experience, it does not confuse the dog; in fact it adds to the powerful message being sent out when he is rewarded for the right behavior. It makes the occasion something special, which in turn gets the dog to try that little bit harder. This practice need usually only continue a short time, until the puppy understands.

D'Arcy reacted well to the routine and soon stopped eating the feces. (This process can be helped, incidentally, by the addition of zucchini or pineapples to the dog's diet. For some reason, both make the feces unpalatable.) Buoyed by this success, I got the family to begin leading him to appropriate places when doing his toilet. Again, I asked them to remain calm and consistent, to keep their pulse rates low. When he went in the wrong spot, they were to simply pick up the droppings and say nothing. They were to do the same if they had missed the moment when he misbehaved. I explained that chastising the dog was even more pointless after the event; the dog would have forgotten about what it had done and would be bewildered by the sudden anger. Again D'Arcy responded to his owners, and within two weeks, he was doing his business in the same spot and leaving it there afterward. The family was overjoyed.

17

The House on Pooh Corner: Soiling in the Home

Despite having been toilet trained correctly as puppies, some grown dogs develop a problem with soiling in later life. While stress in humans shows itself in a vast range of different ways, from physical illness to alcohol abuse, dogs display the problem in their own way. The least pleasant symptom is undoubtedly soiling the home, a problem no dog owner enjoys having to deal with. Over the years I have had to deal with dozens of cases involving dogs that soil their homes; I have been called by people with dogs that urinate when a stranger comes into the house, or that urinate on furniture, on curtains, or even on their owners. It is a deeply distressing problem, and we must once more look to the wild for the explanation.

Wolves and wild dogs are highly territorial species. At liberty in their natural environment, they urinate and defecate to mark the boundaries of their areas. The smells send out a clear signal to other animals: any violation of this space will be met with resistance. It is invariably a job carried out by decision-making dogs, that is the Alpha leaders. It is for this reason, incidentally, that dogs have evolved with the ability to urinate in short bursts. The ability to maintain urine in the bladder allows them to mark out the widest possible area.

While this is the most acceptable behavior in the wild, it is the absolute opposite in the domestic environment. And when a dog instinctively starts soiling, it can be devastating for the owners. Two cases I have been involved in illustrate how the problem can be dealt with quickly and, most importantly, cleanly.

One of the first cases I had to deal with was that of Callie, a Labrador-type mongrel living with a couple in the city of Newcastle. The dog, very much like its owners, Susie and Tom, was a gentle animal. At first she had begun leaving wet patches on the carpet, but her soiling had become much worse. She had now begun to climb onto the couple's sofa where she would urinate freely. The problem had become so acute they had been forced to cover their furniture in rubber sheets.

Like so many of the genuine dog lovers who call me in, Susie and Tom were not angry with their dog. They simply didn't understand what was going on and felt the only way they could help was by learning more about her problem. During our initial phone conversation, the couple had concentrated on the dog's habit of wetting the sofa. People are often so blinded by one overwhelming problem that they fail to see it is connected to many others. So it proved in this case. From talking to Susie and Tom at their home, I discovered that the soiling was far from the only symptom the dog was displaying. Callie was also nervous about going out into the garden on her own, for instance. She would not go out in the dark at all. It was clear to me that this was an example of a dog that was stressed out. And she was stressed out because of the responsibility she had been inadvertently given by her owners.

In this particular case, my efforts to explain the process were made all the easier by the fact that Tom was a fireman. I have often compared the operation of a wolf pack with the way his profession works. The analogy helped him and his wife grasp the principles quickly. Such is the dog's respect for the pack mentality, it will do whatever its job is, at the best of its ability, to maintain the survival of that pack. The prevailing philosophy is "all for one and one for all"; there is no such thing as "I'm all right, Jack." It is precisely the same in a fire crew. In times of danger, they pull together in a manner that we rarely see in our competitive and inherently selfish society. It is, of course, a hierarchical society. Yet from fire chief to new recruit, there is a respect for each other and the community in which they operate. There has to be; each of their lives depends on it. What we had here was a dog who was stressed because it was being asked to do a job it was not equipped to do. I likened it to a situation in which a new recruit to a fire station, a probationer, was being sent out in charge of an entire crew on his first day at work. The couple understood what I was talking about instantly and soon set about applying the Amichien Bonding techniques.

No two cases are ever the same of course. There are often additional routines that owners will need to go through to get success. In this case, in addition to working on the four elements of bonding, I got the couple to concentrate on the "clean dog" techniques I teach owners of new puppies. I encouraged them to follow her around and to reward her when she went to the toilet in the right way. Equally I ensured there were no great dramas if she did not get it right. Calmness and consistency were, as ever, the key. They were not going to relieve the dog of its stress by creating stressful situations themselves.

Even I was surprised at the speed with which it worked. I remember I went to see Susie and Tom on a Saturday afternoon. The following day, Sunday, they rang me to pass on the news that the dog had urinated on the floor. In other circumstances this would, of course, have been dreadful news. In this instance, however, it was real progress. By Wednesday that week, they were on the phone once more telling me she had begun urinating in the designated toilet spot outside the house. She had not soiled the house or the furniture in any way that day.

The ease with which they cured their dog of its problem contrasts sharply with another case, that of a television presenter I got to know during my time on Yorkshire TV. Georgie was a young, attractive, and very lively lady. She doted on her dog, a bichon frise called Derek. Unfortunately, Derek had developed the habit of defecating all over her home. She would return in the evenings to find feces all over her living room. Derek had also got into the habit of doing the same during the night.

As if the problem was not unpleasant enough, the fact that her living room was covered in a dark brownish, rippled carpet meant that she often couldn't see Derek's deposits. Her first job each morning was to lie face down on the floor and scan the floor for any overnight additions. Even this wasn't foolproof. One morning she came down barefoot and stepped right into something Derek had left behind. She confessed to me that she had spent a fortune on rubber gloves and bleach. It was typical of Georgie's humor that she had re-christened the place The House on Pooh Corner. In reality, however, it was no laughing matter.

When I went to the house, the first thing I saw was that Derek followed Georgie everywhere. And whenever she sat down, Georgie would submit to his wishes and pick him up onto her lap. She was, of course, making all the classic mistakes, paying him homage when she was coming in. It was also clear that his soiling was connected to separation anxiety. I learned that Derek was concentrating his activities on the doorway, again marking the entrance to the den.

Like many people, Georgie greeted my method with mild horror. The prospect of withdrawing attention from the dog seemed awful to her. Her natural reaction was to fuss all over him at every available opportunity. I think this was, in part, down to the guilt she felt at leaving him on his own each working day. She felt she had somehow to make things up with him. She quickly saw the benefits of doing things my way, however.

As usual, I entered the home environment displaying all the signals necessary to persuade this dog I was its leader. As a result, after the usual attempts to get my attention, Derek had wandered off and begun entertaining himself, walking off into the kitchen where he started playing with a chewstick. It was only a few minutes later that Georgie realized he had never done this before. I told her that it was because he had sensed I was the leader from my actions and had been able to give up his role as a babysitter. Her job was to convince him in the same way.

We went through the bonding process, again focusing on the techniques I use in teaching puppies toilet manners. I recall imparting another useful tip to Georgie: always use biological washing powder rather than disinfectant when cleaning up after a dog. This is the only way to break down the fatty enzymes within the feces. Otherwise, the dog can still recognize the smell and will almost certainly come back to the same spot to repeat the procedure.

Georgie was, of course, royally fed up with cleaning up after Derek. Unlike the fireman, Tom, and his wife, however, this owner found it hard to adhere to the method. When I saw her two weeks later in the television studio, it was soon obvious she was not following the process properly. Derek was apprehensive in the studio and was looking around at other people rather than to his owner for reassurance. I couldn't help noticing too that she had a pair of rubber gloves in her dressing room. It meant she was not doing the job properly. If she had been, Derek would have been looking to her.

That day, on television with another presenter, we talked about Georgie's problems. Georgie admitted Derek had made huge strides; he would not follow her around as much and he had got out of the habit of soiling overnight. She had not, however, rid him of his daytime soiling habits. I remember she sat there apologizing for failing Derek as a mother!

Afterward, Georgie admitted to me that she was not adhering religiously to the five-minute rule. I had to tell her this was not something she could deal with in a twenty-minute bulletin to Derek each night. My method involved a permanent change in her life and her attitude to her dog. It was clear she had not grasped this.

Because Derek was not getting the message, I asked her to extend the five-minute rule to fifteen minutes. The extra time was needed, less because of the strength of character Derek had been displaying, but more because of Georgie's inability to be strict and therefore convincing as a leader. It was a situation I have come across time and time again: Georgie could not move her affection in a different direction.

In my experience, anyone who truly wants to improve the quality of their life with their dog is, however, capable of overcoming any of the obstacles my method throws up. And so it proved in Georgie's case in the end, I am delighted to say. Two weeks after I had last seen her, Georgie sent me a letter telling me that Derek was a reformed character. She told me she had spent the last fortnight muttering my mantra to herself. She had been calm and consistent with Derek and, as a result, he was going to the toilet in the right place. There had been no more surprises on her carpet. I was overjoyed to get the letter but even happier to see the photograph that accompanied it. It was a snapshot of Derek with his owner's favorite rubber gloves resting by his paws. No longer required around the house, they had become his most treasured toy instead.

18

Situations Vacant: The Problems of Extended Packs

One autumn evening in 1997, I got a phone call from an Irish gentleman called Ernest. Ernest was about to get married but had rung me because he had quite a severe problem, not with his wedding or his bride-to-be I hasten to add, but with his dog. Ernest had known the lady he was marrying, Enid, for more than thirty years. She, like him, was widowed. They had met through their previous partners. The friendship had continued even though Enid lived in the north of England and Ernest was now in Ireland. They had decided to marry and set up home in a new bungalow they were building across the Irish Sea in County Louth.

Their problem was that, while they were looking forward to settling down together, their respective dogs were not. Ernest had bought a cross-bred bitch puppy called Gypsy shortly after his first wife died. In the seven years since then, Gypsy had become the absolute idol of his life. Similarly Enid had a deep affection for her dog, a thirteen-year-old Labrador-cross called Kerry. Ernest had begun visiting Enid in her home each month and had tried to introduce Gypsy to Kerry, but neither dog was having any of it. The couple had tried everything, including an animal behaviorist who had produced a long five-page report but who had done nothing to actually improve relations between the two dogs. They were very depressed.

I arranged to meet the couple and their dogs at a friend's boarding kennels and, firstly, decided to take them all for a walk. It was soon pretty obvious that the two dogs were eyeing each other up—if I am honest, Gypsy more than Kerry. There was definitely a strained relationship there.

Everybody who calls me out cares about their dog enough to want to sort out its problem, whatever it is. They don't just put the dog down because of its biting, or put them into a home because they can't cope with them. Ernest and Enid were so determined to solve the problem, they were willing for me to take charge. The problem here was that Kerry was protective of Enid and Gypsy was protective of Ernest. Both dogs perceived themselves as leaders within their individual packs. They were now challenging each other for the vacant position within the new, extended pack. What I wanted to do was make the two dogs dependent on each other for comfort and companionship, to form a pack of their own. I would then set about establishing them as equal subordinates in that pack.

The first thing I asked the couple to do was to leave both dogs at the kennels near Enid's home. For a couple of days we put them in kennels side by side, so while they were deprived of their beloved owners, they did sense each other's presence. On the third day, I went there and took them into the large exercise area. The reason was that I wanted the dogs to have room to separate from each other but at the same time be in a common environment. They both had their comfort space.

The dogs gave each other quite a wide berth; they were pretty dismissive of each other. For me this was cause for a lot of hope. I did this for three days running, and by the third day, they seemed to want to get to know each other. They were wagging their tails, offering playful gestures to each other. This was the sign I needed to take them onto the next step. The next day I put them into the same kennel. There were two beds, two bowls—everything was separate if they wanted it to be that way—and there was plenty of space too as, it was a large, double kennel. That evening I had a phone call from my friend who ran the kennel. She told me that one of the beds was already redundant as the dogs were sharing. I was delighted.

I resisted the temptation to tell Enid that it was going well because there is nothing worse than building up people's hopes and for something then to go wrong. Instead I went on to try the next step. We left the dogs like this for a good week, during which time they were coming on nicely.

With Ernest in Ireland, I asked Enid to come to the kennels first. The important job now was to establish both dogs below the two owners in the pecking order of the extended pack, to show them it was pointless jockeying for the job of leader because the role was not vacant. I asked Enid to totally ignore them when she saw them. My reasoning was that Kerry would automatically think, "This is my baby, let's have fun," and that Gypsy would feel left out. I wanted her to leave them both feeling left out, so that they turned again to each other. We had a nice session of about half an hour, during which time Enid showed the dogs no affection whatsoever; she did not stroke either of them, she did not even make eye contact. This may seem very hard but I wanted to establish to the dogs that there was no challenge to each other while Enid was around. We did this several times and, each time, Enid got slowly more friendly with the dogs, stroking them, giving them rewards but always very quietly. She knew calmness and consistency would be the key to everything we were doing.

On his next visit over, I asked Ernest to repeat what Enid had done. I wanted him to do it alone, as Enid had done. When Gypsy saw him, she got very, very excited indeed. She grumbled at Kerry more than once. Had Ernest made a fuss over her at that moment, it is quite possible that Gypsy would have become quite aggressive toward Kerry, which was the last thing we wanted. Again, Ernest was determined, and even though it was difficult, he did it. We repeated the process again during the two days he was over, very successfully.

Before Ernest went back to Ireland I decided we could try one run together, all five of us. The great day arrived and we were standing in the exercise area relaxed and happy. I can't begin to tell you the joy I felt at this time because these people had put their faith in me to do something that would quite dramatically change their lives for the better. And it was working.

Shortly afterward, I was invited to Enid and Ernest's wedding. After the service, I was to my surprise invited back to the reception and, as we went through to the dining room, they pointed to a seat for me at the top table. Ernest began his speech by thanking me for all that I had done for them. I felt overwhelmed to say the least. It was then that it hit me what this process could mean to people. It was one of the most humbling experiences of my life. I knew that for them to be really contented in their lives, these dogs that they loved would have to get on. I didn't realize how much it meant to them until that day.

The following week, it was arranged that the dogs would go over and join Ernest and Enid in their new home. There were a few phone calls but only minor problems. The new family settled down together wonderfully. It was about a month later that I got a phone call from a distraught Enid. She told me they had been into Dublin shopping that day and that somehow Kerry had got out of the car and got lost. She had disappeared into the streets. Enid and Ernest had been to the police station, made an appeal on the radio station, put posters on the streets, everything, all to no avail. They were devastated, as I was for them.

After ten days they had virtually given up, then they got a phone call from someone in Dublin who had found a stray dog fitting their description and had taken it in. They went back up there in their car and sure enough it was Kerry. Enid thought they were pleased to see her. But what really moved them was that Kerry shot past them and raced straight to the car where Gypsy was waiting. When they opened the door, Gypsy leapt out, crying and twisting in the air with absolute delight to see her friend. I still get Christmas cards from the four of them—"Ernest, Enid and the girls"—and whenever I do, I picture that moment.

19

Biting the Hand That Feeds: Problem Eaters

On the face of it, feeding time should be the most straightforward part of a dog's daily life. Eating is the most elemental of instincts after all. It is surely a simple matter of placing a bowl of food on the floor and leaving the dog to it. Isn't it? Well, yes and no. Provided there are rules under which food is being dished out, meal times should not be a problem. The difficulty, as I have discovered in a number of cases, is that dogs have a habit of dictating those rules themselves. And that is a recipe for nothing but anarchy.

Of all the dogs I have dealt with, the most interesting was an eleven-month-old Tibetan lhasa apso called Jamie. Jamie had arrived in his owner's home at the age of eight weeks and had always been a picky eater. At some point, the family had begun feeding him by hand. In the month before I was called in, however, his eating habits had declined to almost nothing. He resolutely refused to eat anything his owners put in front of him. As they grew more desperate, they had tried everything, from prime steak to the most expensive prepared pet foods. They even ordered him a meal from the local Chinese takeaway in the hope that it might appeal to his ancient Oriental heritage! All to no avail. He was now painfully thin and his ribs were beginning to poke through. Adding to their frustration was the fact that he would prowl around the bowl endlessly but never dig into it. Jamie had been taken to a vet who had found nothing wrong physically. It was the vet who recommended his owners should call me.

As I have explained already, it had been while observing life within wolf packs that I first came to see the immensely important part played by food. One particular incident I saw in a documentary film always springs to mind. The film observed a coyote as it circled the carcass of an elk which had been slaughtered and eaten by a wolf pack. The wolves were resting after having eaten their fill, about three-quarters of the carcass. The outsider's presence was clearly not appreciated, however, and it was the Alpha female who drove it away. What was interesting was what happened afterward. After seeing off the coyote, the Alpha female returned to the carcass and almost ritualistically pulled off a piece of meat. The message to the coyote was clear. She had the power to decide who should eat and when. She was reasserting her leadership in the most powerful language imaginable.

I have seen this behavior replicated almost precisely in dogs. Many an owner has smiled sweetly recounting how their dog regularly appears with a biscuit in its mouth. Part of them is, I'm sure, disappointed when they learn the dog has not come to display the biscuit because it is hungry but rather to reassert its place as the household's prime distributor of food.

When I traveled to meet Jamie and his owners, it was soon clear his behavior was rooted here too. As soon as I arrived in the house, I saw the classic signs of a dog that believed it ruled the roost. He jumped around and barked furiously when I arrived, clearly keen to put me in my place. I, of course, ignored him. When I sat down with his owners, he jumped up on their laps, sitting in on the meeting. I was not at all surprised when I discovered a supply of food sitting in a bowl in the corner of the kitchen. Similarly I barely raised an eyebrow when Jamie's owners told me it remained there twenty-four hours a day and was replaced with fresh supplies at least three times daily. It was clear to me that food had a particular significance to Jamie. Just to be one hundred percent sure, however, I headed to the bowl. The moment I did so he scampered over barking even more furiously.

I explained to the owners what was happening. The reason he was not eating from the bowl had nothing to do with his lack of appetite. All dogs react differently to finding themselves out of their depth in the role of leadership. This little puppy's reaction had been to fixate on food, what he saw as the ultimate symbol of his power. This was why he patrolled it like a guard on duty at Fort Knox, almost daring his owners to eat from it. And that was why he never ate from the bowl. On the surface, it is completely irrational. His action was ultimately destroying him. And I have no doubt that this little dog really would have starved himself to death. But why should a dog behave according to the logic of another species? Seen from this perspective, it all made perfect sense to his owners. Why would any leader eat the contents of his power base?

The family's treatment of the problem had been the polar opposite of what was required. Of course, I understood completely why the family had done what they had done in placing food all over the house. And it was clear to me that their decision to feed Jamie by hand had been more responsible than anything in starting the decline. To the dog, that would have seemed like groveling of the highest order. It would only have added to his belief that his pack was totally reliant on him. My job was to explain to the family the need to shift the balance of power within the household—and the power of feeding time in particular. I requested that the family apply the usual bonding techniques. But in this case, I also asked them to concentrate their attentions on meal time, carefully going through the gesture-eating routine three times each day. However, if Jamie deserted the bowl, they were to pick it up and not put it down again until the next time he was due to be fed. This gave Jamie no choice: either eat when he was provided for, or go hungry.

Jamie's stomach was by now shrunken so I asked them to feed him only morsels of food. He was, in addition of course, going to be offered lots of rewards for his actions as part of the rest of the process. On the first day, he hardly ate anything, partly due to his delicate condition but also because his owners were giving him signals he had never seen before but which he understood. He needed time to think. By the second day, however, he had got the message and was eating again. He ate two mouthfuls from his first meal, three more from his second. To the family's delight, he ate his entire dinner that evening. By the fifth day, he was eating three full meals a day. By the time his first birthday arrived, he was back to his optimum weight and displaying all the signs of being a normal, well-adjusted little dog.

Jamie's problems were far from uncommon in a puppy. Eating time has the potential to send out more false information than almost any other situation. It is why it is one of the key elements of my method. The wrong signals can prove disastrous. And the younger and more impressionable a dog, the greater the scale of that disaster can be. It is no surprise many people get this wrong. There is, I am bound to say, a lot of confusing and downright dangerous advice around regarding food. For instance, I have in the past seen supposed experts argue that it is good practice to remove the food from the dog while it is eating. One television program I saw, filmed at one of the best-known dog homes in the United Kingdom, showed trainers bringing a dog into a room on a lead, placing it in front of a bowl of food, then doing their level best to remove the bowl from the animal as it ate. The more they tried to disrupt its eating time, the more the dog growled and snapped at them. As a result of its actions in this situation, the dog was destroyed.

In my opinion, those so-called experts killed a dog for no good reason at all. As I have explained, feeding time is absolutely sacrosanct within the dog's natural environment. Every dog takes its turn. And during its turn, nothing can be allowed to interrupt it. I can think of nothing more certain to provoke a dog into defending itself than trying to interrupt its feeding time. The dog home's argument—that if they could not remove the food, the dog must be too dangerous to be given a new home—was unfair. I must admit I wept when I watched what they were doing.

I have witnessed the type of aggression this poor dog displayed many times. None proved how effective my method is in solving the problem more than Mulder, a golden cocker spaniel. Mulder had a perfectly good appetite. The problem his family had was that he was just too aggressive and impatient in his desire to take charge of his meal times. Whenever meal time arrived, Mulder would start growling. As Yvonne opened his tin of food, he would get more and more aggressive. Worst of all, he had developed the habit of jumping up and biting Yvonne's hand as she placed his bowl on the kitchen floor, a classic case of biting the hand that is about to feed you if ever I saw one. To Mulder, the Alpha, it did not make sense that a subordinate was feeding him; any dog owner who has had a dead animal brought to them by their pet will have witnessed a dog trying to reverse these roles. In Mulder's eyes, Yvonne was behaving badly by having access to the food before he did.

When I arrived at the house, my job was to show Yvonne how she must handle feeding time from now on, so I explained the process of gesture eating. Mulder had, of course, been called after the character in television's *The X Files*. I'm sure Yvonne never found the show as scary as her dog. Her nerves had been so shredded by Mulder that she was shaking violently as she walked into the kitchen. Somehow Yvonne regained her composure, prepared a cracker for herself, then emptied Mulder's meal into his bowl and placed the two together on a raised surface. Mulder's expression was frozen as Yvonne began eating first. He couldn't quite believe her audacity. I stressed to her that she must take her time. This she did, chewing away for a full minute or so while her dog continued to look on in disbelief.

Only when she had made a fuss of showing she had finished did Mulder get his meal. She was so terrified that she had taken to throwing the food on the floor. So as to reassure her, I placed his bowl on the floor without making any sound at all, then left him to it. Gesture eating communicates one of the most powerful messages available in the language of a dog. It never spoke more loudly than it did in Mulder's case. To borrow a phrase from *The X Files,* the truth was out there. Yvonne simply had not known where to find it. After two weeks of the process, Yvonne was preparing Mulder's meals in peace. He has not been a problem since.

20

Have Dog, Won't Travel: Dealing with Car Chaos

To many dogs, the back seat of a car can seem like hell on earth. In the course of my work, I have come across a dog that barked the 200 miles and four hours from Lincolnshire to Scotland and another that would literally try to scramble out of the window on motorways. Many owners had admitted defeat and given up on the idea of traveling more than a few miles with their petrified pet.

Yet a dog's anxiety is hardly surprising if we think about it. In almost every respect, the car is little more than a condensed version of the den. Whenever it gets in there, it is surrounded by some or all of its pack. Yet coming at it from all angles are an array of sights and sounds it does not comprehend, cannot reach, and is convinced is going to harm its charges. Placed in such a situation, who wouldn't go into a blind panic? The reality is, however, that any owner can deal with the problems of what I call car chaos. Two cases I have encountered illustrate how easily and effectively even the most severely disturbed dogs can be transformed into happy travelers.

The Cleethorpes couple who owned Blackie, a Labrador-Border collie crossbreed, had tried everything to overcome his habit of going ballistic the moment he was placed in the back of their car. They had tried turning the radio up to full blast, they had tried shouting at the dog. Nothing was working. Every journey had turned into a nightmare—even the daily half-mile run to their local beach, where Blackie would then proceed to enjoy his walk.

I spent the first hour or so of my visit in the normal way. As I explained my method to Blackie's owners, I simultaneously began bombarding him with the signals central to it. As Blackie began to disregard his owners, he was soon coming to me well. When people first see their dog connecting with me in this way, they are often concerned. They wonder whether I have somehow directed the dog's affection away from them, if I have somehow taken over from them. The reality, of course, is that the dog has found a leader that it believes can look after every member of its pack. It is a process that they will then have to go through themselves. They soon see that the best way for me to illustrate the method's power is to carry it through myself. Their bond with the dog remains the same, it is just the power base that alters.

Soon I felt I had made enough progress with Blackie to attempt going out on a drive with him and his owners. As we climbed into their car, they took their normal position in the front and Blackie took his in the back section of the estate car. I positioned myself between them in the rear passenger seat. Unlike so many people who—quite wrongly in my view—let their dogs roam free in the car, Blackie's owners had confined him behind a guard at the back of their estate car. I kept him on a lead and passed the lead through the guard so that I had some control over him.

As the engine started, I sat as quietly and calmly as possible. As we moved off, I placed an arm back through the rails of the guard and placed it on Blackie's shoulder. When Blackie began to attempt to jump up, I applied a little more gentle pressure. He immediately eased back down.

We traveled for about three or four miles, heading, quite deliberately, through the busiest part of the town. I wanted Blackie to be faced with as many sights, sounds—and to his mind potential threats—as possible. Throughout the journey, I kept my arm in place on his shoulder. Each time he showed any sign of being apprehensive or excited, I gently increased the pressure. There is, in these instances, a fine line between force and reassurance. Most people can understand this instinctively. For those who can't, I compare it to holding a child through its first visit to the dentist. It is a painful but necessary process. By ensuring the child is sitting calmly, it will be that much less traumatic. By the time we returned home, I hardly needed to keep my arm there at all. Blackie had spent the latter part of the journey simply sitting in the rear of the car, watching the world go by. He has been happily traveling in the car on a daily basis ever since.

Like humans, dogs can carry the scars of previous experiences. Anyone who has been involved in a car accident, for instance, finds it hard to get back into the car afterward. It is no different with dogs, as I discovered when I was called in to deal with a particularly distressing case. The experience this particular Doberman had suffered had been so horrendous, it had made the pages of a local newspaper. He had been found wounded and deeply distressed on the verge of a motorway. It is barely believable, but what appeared to have happened was that his owners had physically thrown him out of a speeding car. The dog's injuries were so horrendous it had been confined to an Intensive Care Unit. At one point, it had not been expected to live. Slowly but surely, however, it recovered. He was eventually taken in by a couple in the village of Barnetby. They quickly saw that one major mental block remained in place, however.

Dobermans are no shrinking violets, yet the merest sight of a car was enough to send him into a panic. When his owners had managed to force him into the car, he had duly urinated all over its interior. It would have been all too easy to write this dog off as a lost cause, his trauma was so severe. Again, however, I was dealing with people who genuinely cared for the dog's welfare. They were determined to try everything possible.

During the day I spent with them, I explained they faced a long haul. This was a dog that was going to need a great deal of reassurance before he willingly went anywhere near a car again. Fortunately, they were excellent learners. After two weeks or so, they had established leadership in the normal way. I then asked them to focus as much activity as they could on and around their car.

So began another month of work. They began by placing a bowl of food in the driveway with the car in full view. The idea here was for the dog to rid itself of the idea that the car was something with a purely negative association. From here I got them to move closer and closer to the car. Again, I stressed the importance of calm and consistency here. They took their time, even starting to eat their evening meals on deck chairs in the driveway so as to underline the message they wanted to transmit. Eventually their work paid off. The breakthrough came when they persuaded him to eat his dinner in the back of the stationary car. From there they began playing retrieval games with toys in and out of the car.

Progress was painstakingly slow, but the owners were determined to make it work. Soon they had progressed to switching the engine on while he ate in the back. Then they moved up and down their driveway while the dog ate his meal. The mental scars were so bad it took almost eight weeks for them to get the car out onto the open road. I am glad to say, however, that they are now driving around as a family quite freely. His fear of traveling is a thing of the past.

21

Feet-Chewers and Tail-Chasers: Nervous Wrecks and How to Salvage Them

All dogs have different characters. Like humans, there are those that are playful and those that are placid, there are the extroverts and the introverts. It is why individual dogs deal differently with the stress they encounter when they are given the job of leader. While some will lash out at the world, others will turn in on themselves, often in the most self-destructive ways. In the course of my time dealing with problem dogs, I have seen a range of symptoms that beggars believe.

I have come across dogs who cower from the slightest, most innocent noises. A gentle ring of the telephone is enough to send them scampering for cover. Some dogs are so chronically nervous I consider it a major achievement if by the end of my time with them they have come more than two or three feet toward me. I have seen dogs that freeze at the sight of anyone dressed in uniform. On many occasions I have seen dogs who display the ultimate signal of submission and simply flatten themselves out on their belly and wet themselves. I am sure I will continue to come across new manifestations of this problem for as long as I continue to work with dogs. The root cause of this behavior is always the same, however. The dog is simply overwhelmed by its responsibility as leader. It manifests this through nervousness and obsessive behavior.

Riby was a four-year-old black Labrador named after the village near Grimsby in which he lived. His owners called me in because Riby had developed the particularly nasty habit of chewing his own feet. The problem had started as a minor habit but had grown more and more obsessive. When they called me, he was at the stage where he would bite at them continually. Obviously this was far from healthy and Riby had a series of nasty open wounds. If it continued like this, there was every prospect that his feet would become infected, perhaps even gangrenous and he would be destroyed. His owners were understandably desperate to find a solution. They had tried all manner of treatments, including sedation. When I visited them, Riby was wearing what I call an "Elizabethan collar," a white, plastic funnel over his head. The collar was designed to stop him from getting at his feet.

Riby displayed the normal range of symptoms. So many people imagine that it is normal for a dog to jump up, to pull on the lead, to harass visitors to the house. I can assure them it is not. Riby did all of these things too. Most telling of all, he had got into the habit of lying in his bed in the morning. He would not get out until his owners coaxed him out. It was as if he was lying in state; it was the powerful sign that once more I was dealing with a dog that believed it was leader.

I began by going through the normal bonding process. Riby responded well. I sensed fairly quickly that this was a timid dog that was ready to relinquish its leadership as soon as possible. After about one-and-a-half hours, I asked his owners to take off the collar. No sooner had they done this than he had started gnawing at his feet. Riby's problem was a variation on the human condition known as self-mutilation. The important thing was to show Riby there was no need for him to do this—that he could be rewarded for other activity.

So I knelt down and called him to me with a reward. When he came, I covered his feet with my left hand and with my right hand cupped his head and stroked his chin. I did this without saying a word. I wanted the process to be nonstressful, calm. He remained distracted for a few moments but soon started picking at his feet again. As soon as he did, I again distracted him. This time I asked him to come to heel, again rewarding him with food. Again, it was a positive association. I continued in this way for a while. Every time we stopped and he turned on his feet, I started working with him again. I just kept him going. We worked like this for about twenty minutes. He was behaving much better by this time so I went to the kitchen for a cup of tea with his owner. As we chatted, we forgot about Riby for a few moments. It was a few minutes later when we noticed Riby had fallen fast asleep on the living room floor. At last he had given up the stressful role of guardian and could relax for the first time.

This was the first time I had ever come across such severe behavior so I asked his owner to keep me posted on his progress over the next few days. I think I heard from her once or twice over the coming weeks. Her message was the same on both occasions: Riby's feet had healed and he had adjusted back to normal life. In the aftermath of our few hours together, he had never eaten his feet again.

The psychology of the dog is a subject for another book—and a rather large one at that. I am not going to analyze the workings of the canine mind here. All I will say, however, is that the dog has a capacity for obsession that is no different from our own. I have seen all manner of weird and wonderful behavior over the years. A German shepherd called Rusty, for instance, could spend entire hours chasing his own tail. His owners could not fathom what he was doing and called me in. I arrived to discover a fairly well-adjusted dog with a few telltale signs of leadership. He jumped up and whined a little, but not in an excessive way at all.

It may well have taken me some time to work out what

was causing the problem. Lady Luck was smiling on me that afternoon, however. As I spoke to Rusty's owners, their three-year-old daughter fell asleep. Rusty was clearly deeply attached to the little girl and duly curled up alongside her. The girl did not sleep for very long at all. It was as she woke up that the lights came on in my head. As she came around, the little girl instinctively reached out for Rusty's tail. She took hold of it and began to shake the tip of it around like a plaything. Almost immediately Rusty was transformed into a whirling dervish. He was up off the ground, spinning around.

His owners had never noticed what was happening before. I explained this was being caused by their daughter's playing with the tail. As I have said earlier, it can be difficult to teach a young child the correct way to behave around a dog. In this instance, I asked the parents to keep the two apart when unsupervised. I also asked them to play games which focused the little girl's attention away from the tail. I got them to play retrieving games, anything that got the girl to concentrate on the dog's head area rather than its rear. Soon, Rusty had been relieved of his habit. His whirling dervish routine disappeared, and he was free to spend his life chasing toys in the park instead.

22

The Yo-Yo Effect: Overcoming the Problems of Rescue Dogs

Animal sanctuaries and dogs' homes have become, for many people at least, an ideal place to find a new pet. The idea of taking in a dog that has had a hard time in life appeals on many levels, of course. It is heart-warming for dog lovers to think they might be able to provide a little of the affection that has been so sorely missing in the lives of these canine waifs and strays. If they are taking on a dog that has misbehaved in the past, they like to think they are the ones who can straighten it out. Yet the rescue dog comes with its own unique set of problems. More often than not, in my experience, the behavior that has led to its being abandoned or consigned to a home in the first place repeats itself time and again. And when this happens, owners who start with the best of intentions find themselves unable to cope. It is why so many dogs become what I call "yo-yo dogs," spending their lives heading back and forth between families and institutional homes. In the end, of course, they run out of chances and may even face being destroyed. Only by understanding their particular problems can owners hope to avoid this and provide a happy and permanent home.

The first thing to say here is that it is not the dog's fault that it has become caught up in this vicious circle. In 99.9 percent of cases, the dog's behavior is a direct result of human mistakes, mainly laziness, stupidity, or, sad to say, cruelty. The problems displayed by almost all rescue dogs have been exacerbated by the violence they have faced at some point in their life. Violence begets nothing but more violence, however. The irony is that dogs that have been confined to a home for attacking humans have merely been defending themselves. They have generally been cornered in situations where the option to flee has been removed. Within the human world, self-defense is a perfectly acceptable legal principle. For dogs, however, it is they who have to bear the consequences, regardless of where the blame lies.

I saw first hand the traumatic effect bad treatment can have on a dog when I took in my own rescue dog, Barmie, the little companion who taught me so much when I was evolving my method. If there was one central lesson I learned in working with him, it was that the bond of trust between dog and owner is even more important in cases like this. Barmie, quite rightly, harbored a deep distrust of all humans. He—like all rescue dogs—needed to learn that the hands that had brought pain can deliver affection and food too.

As in medicine, prevention is far better than cure. During the course of my television series, I was asked to help prepare an owner for the arrival of a particularly troubled little dog. Tara had been taken in by a friend of mine, Brian, who ran a refuge in Leeds. He had discovered she was within a day of being destroyed. What made her case even more heart-rending was the fact that she was heavily pregnant at the time; her puppies would have perished too. Brian had seen Tara through the birthing process and was now ready to find her a good owner. In Hilary he had found the perfect owner. She was a genuine dog lover, desperate for a new dog with which to share her life.

As is so often the case with rescue dogs, there were no real clues as to why Tara had been abandoned. She had behaved perfectly well at the sanctuary and seemed a normal, well-adjusted dog. I tell people not to worry about a dog's history. The past colors everything a dog does but it is rare that anyone can supply a full history in any case. Far better, I think, is to concentrate on the dog's future instead.

Of course, Hilary wanted to do everything she possibly could for this poor dog. She had, for instance, laid down some food for her arrival. After I explained why this was not appropriate, she removed it. In my experience, it generally takes two weeks for things to go badly awry. It is at this time that the dog is transformed from a lovely, peaceful dog to one that seems completely out of whack with the rest of the world. In Tara's case, it took even less time than I had imagined.

At first Tara just mooched around. Hilary was so eager to fuss over her that I had to keep telling her to leave her alone. After a short while, however, Tara approached her new owner directly. She came over to Hilary and placed her head in her hands. It was here that Hilary made her big mistake.

Hilary instinctively stroked her new companion. Truth be told, she had been aching to extend an affectionate hand in Tara's direction ever since she had arrived. It was the trigger Tara had been waiting for. Tara immediately jumped and leapt around. She became completely hyperactive. It was as if Hilary had tripped a switch inside the dog's head. And it was as if Tara really was a schizophrenic, as if she was two dogs rolled into one. It was soon glaringly obvious why she was in care. In a home situation like this, she became completely hyperactive. A succession of owners had been unable to handle her behavior. Her nomadic existence had been the result.

Hilary was determined to break the cycle, however, and was set on understanding what the problem was. I had already outlined the general principles of my method. As we watched Tara flying around the house, I explained that the normal problems ran much deeper than usual because of this dog's history. As I have explained in detail earlier, dogs can become highly stressed by the role of leader. In the case of a rescue dog, the pressure is almost intolerable because the stakes are even higher. We only have to think about it for a moment to understand. Here is a dog that desperately wants to be a part of a normal pack environment. Yet as soon as it finds a home it likes, it is elevated to the role of leadership. When the dog finds it cannot cope with this responsibility, it tries even harder to please its owner. When the owner reacts violently or angrily, its behavior becomes more and more excessive. I have seen countless cases where rescue dogs jump up, drag on the lead, bark and bite, and become generally hyperactive. They genuinely believe it is what their subordinates expect of them. In more ways than one, it is a vicious circle. The owner's reaction only serves to stir the dog into an even greater frenzy. And soon the dog is being returned to the sanctuary from where it came, its reputation as a problem dog underlined by what has happened. The yo-yo effect has continued.

I explained to Hilary that the answer lay in treating the central problem rather than the symptoms. Tara had to be taught that, in fact, this was completely the wrong way to behave in a household. The job Hilary faced was to introduce a different set of rules. As ever, I stressed the importance of good, strong leadership. I asked Hilary to maintain a quiet posture and ignore the command performance Tara was turning in. Everything was telling me that in the past the reaction had been the exact opposite of this. Every time Hilary looked like weakening, I reminded her of what lay in store for Tara if we failed.

Sure enough, Tara had soon calmed down. There were a few, inevitable attempts to reel us in to her world. She kept trying to make eye contact with Hilary but failed. After a while, she went to lie on the floor. Once she was completely relaxed, I told Hilary to wait five minutes. At the end of that time, Hilary called Tara to her with a reward. Tara didn't get it right immediately and started jumping around again. Once more, I told Hilary to step away and ignore her. Only when Tara played precisely by Hilary's rules did she get the reward. It was up to us to show her how she should behave. Within half an hour, Tara was a different dog. She and Hilary have been fantastic friends ever since. The cycle had been broken; she was a yo-yo dog no more.

23

Toys Not Trophies: The Power of Play

I would not want to create the impression that all my ideas are unique, that I have come up with a complete process of techniques unlike anything ever used before. As I explained at the beginning of this book, I drew many of my early ideas from behaviorism. In many ways, I am heartened when I see elements of my work incorporated elsewhere. I was never more surprised to see an element of my method being practiced than in the spring of 1998, when I was invited to visit the country's biggest and most high-profile dog training facility, the London Metropolitan Police's handlers' school at Bromley in Kent.

I joined in a session with a senior trainer called Eric, in which a group of German shepherds was being taught to force humans out of situations in which they are hiding. There were fascinating elements to what the police were teaching them. The dogs, for instance, were coached to remain at least six feet away from a target. Eric explained this was a simple matter of survival; any closer and they would be prone to an attack by kicking or, even worse, a knife.

It was within this highly charged, very serious situation that Eric did something that put a knowing smile on my face. The object of the exercise was to encourage the dog to bark so furiously that it intimidated and ultimately forced the surrender of the human. Sure enough, the first dog pinned us into a corner with the sheer ferocity of its demeanor. When he was happy that the dog had done what was asked of it, Eric reached into the collar of his jacket (the dogs had been taught to react to any body movement lower than this). From there he produced nothing more sinister than the dog's favorite toy, a battered old rubber ball. When he threw the ball over the dog's shoulder, the terrifying animal of a moment earlier was transformed into a bouncing, bounding child. The dog's handler had, of course, taught the dog to respond to the ball in this way at the very beginning of its training. Ever since, it had remained a very powerful means of signaling that the dog had done something that met with his approval. It was a means of reward, and one that I recognized very well indeed: play.

Playtime provides perhaps the perfect opportunity to combine fun with learning; there is no greater pleasure. Yet it is precisely because it holds such a potent place in the relationship between humans and dogs that play has to be conducted in the right way. It may not seem like a particularly severe problem, yet being dictated to in this way can have disastrous consequences. I'm sure we have all found ourselves in a situation where we have just settled down at the end of a hard day only to see our dog appear with a plaintive expression on its face and a favorite toy dangling from its mouth. The dog wants to play, and it wants to play now. Hard as it is for most owners to see this at first, the situation is filled with potential problems.

The act of throwing around a ball or retrieving it should be viewed from two perspectives. To us, these objects are mere toys. To the dog, however, they represent something far more precious. They are trophies, badges of honor if you like, to be won—and lost—within the pack environment. Packs of puppies, in particular, fight for objects continuously. The winners strut around as if they've just won the World Cup.

Again, it is a principle that extends all the way back to the wolf pack. In the wild, the pack's survival depends on its leaders being up to the job. As a result, the Alpha pair must regularly prove themselves worthy leaders. Dogs constantly test the mettle of their leaders in the same way, and playtime offers a perfect opportunity to do this. If dogs are allowed to believe they have control of the toy "trophies" an owner throws, they will also form a belief about their status within their pack. Therefore, it is imperative the owner imposes himself as the leader during this time.

Problems set in when the owner refuses to join in the game. Much like a child who throws a tantrum when denied something, a dog can treat the lack of response with bad behavior. I have known cases where pets have started a nightly routine of getting into an agitated state over toys. It can escalate into destructive and even aggressive behavior.

There are a few simple rules I apply at playtime. The first and most powerful means of establishing control over playtime is the simplest. I discourage owners from leaving all the dog's toys around the house. It is good practice to have one or two favorite toys available. That way the dog can choose to play on its own whenever it wants. But it is essential that the toys with which the owner interacts with the dog are stored away in a place where the dog cannot get to them. That way the power of playtime is entirely in the owner's hands from the very beginning. It is they and they alone who decide when playtime takes place and which toys are used. As for the choice of toys, that is entirely a matter for the owner. The only caveat I would add is that all toys should be of a decent size. As with puppies, dogs can choke on balls, for instance, that are small enough to squeeze down their windpipes.

As for playtime itself, one of the golden rules I emphasize is that an owner should never get into a tugging contest with a dog. There are two very good reasons for this. Firstly, it allows the dog to dictate the rules of the game. Secondly, and potentially even more dangerous, there is a danger the dog may sense its physical superiority over an owner. And if it begins to believe it is stronger than its leader, it will begin to reassess whether the leadership should rest with its human companion any longer.

I frequently use playtime to practice and top up some of the key disciplines with my dogs. Skills like the recall and coming to heel need regular refreshing. By moving away from my dogs when they retrieve and return to me with a ball, I encourage them to come to me. They want the game to continue. They know that for it to do so, the ball must be back in my hands; they want to carry on playing, so they behave in a way that ensures this happens.

I have been asked to deal with all manner of problems in this area. None was quite as interesting as that of Benji, a lovely "Westie" or West Highland terrier. Benji had a unique problem. His owner, Mavis, called to tell me he was behaving very oddly whenever she brought out a new squeaky ball. Benji had always enjoyed his playtime, and enjoyed playing with squeaky balls in particular. The sight of this new ball would transform him, however. Sure enough when I visited Mavis, I saw the reaction first hand. He lay down, placed his head on the floor, and just trembled.

It did not take me long to work out what the problem was. Mavis told me that Benji had always punctured any squeaking toy within minutes of being given it. This larger one had remained intact, however, because he had not been able to get his jaws around it. Terriers in particular are known for their abilities as rat catchers. I suspected that Benji's habit of puncturing balls so that they could no longer squeak was connected to this. He had, in effect, failed to kill the King Rat that was his giant ball. This had left him feeling terrified.

I knelt down next to Benji and, making sure he saw what I

was doing, drove a screwdriver into the ball. He watched attentively as I made sure all the air had been expelled and the squeaking noise eliminated. His reaction was unbelievable. The second the squeaking had been removed, Benji grabbed the ball, tossed it off the ground, and started leaping in the air with it. His ears were up, his whole body was trembling again, but this time with excitement. His mortal enemy was no more. When I threw the ball to him again, he ran around with it triumphantly. It remained his favorite toy for months afterward.

24

"How've Ya Done That, Lady?"

Since first developing my ideas, I have become more and more secure in my belief that man and dog form a unique relationship. Each time I see a newspaper or scientific magazine produce new evidence to support this, I feel more positive that the powerful means of communicating I use is somehow reconnecting us with our past.

The more I work with different breeds and particular problems, the more my ideas have unified around the method I have outlined in the previous pages of this book. It is, much like our relationship with the dog, an ever-evolving process. People often refer to me as an expert. I always reply the same way: it is the dog that is the expert, I am only someone who has learned to listen to it and now feels ready to share what I have heard with others.

As I have done so, I hope I have helped many people learn to train and live with their pets in a compassionate way. There have, inevitably, been instances when my efforts have not been sufficient. Ultimately, it is up to every individual owner to put the principles of my method into operation; it is not a quick fix to be forgotten, it is a way of living with their pet. A few—fortunately very few—owners have failed to grasp this and their dogs have suffered the consequences.

In the vast majority of cases, however, I have been able to help. And as my method has gained more credence, I have found myself being able to help in increasingly emotive situations. I have on a number of occasions now been asked to intervene in the cases of dogs facing the legal threat of destruction. One such case was that of Dylan, an Akita.

Dylan's owner was a saleswoman called Helen. As she traveled the length and breadth of the country, Helen took Dylan with her. He acted as companion-cum-protector. And, given the fearsome power of the Akita breed, he fulfilled the second role with ease. Unfortunately, his protective streak proved too powerful.

One day Helen was loading some shopping into the trunk of her car in the car park of her local supermarket when she was approached by an acquaintance. The door of the car was opened. As Dylan saw the lady extend her hand to Helen, he leapt into action. The wounds to the lady's arm were so severe, she needed hospitalization and many stitches. The attack was so severe, the police were involved and Dylan and Helen were prosecuted under the Dangerous Dogs Act. A judge was to decide whether Dylan was to be destroyed or not.

Helen contacted me via her solicitors. She did so for two reasons. Firstly, of course, she wanted if possible to save her dog. But more importantly, she was determined to find out why her dog had done this. Of course, the two things were linked. If she could solve the riddle and change the dog's ways then the court might look on it more sympathetically.

Her amazement was obvious when she first called me. "I don't understand why he did it," she kept saying to me, "he is so lovable." As ever, Helen was unaware of the other symptoms Dylan had been displaying. When I asked whether he had been following her around the house, whether he became agitated when visitors came, and whether he tended to protect her, she replied yes in every instance.

I told Helen she must be absolutely diligent in using my method; the dangers of applying it inconsistently had been proven by the case of another Akita I had dealt with. Despite my requests, the owner in that instance had not applied my signals consistently and the dog could not improve. When he bit again, although the courts were not involved this time, he had to be destroyed nevertheless. His owners were understandably devastated.

Helen had two months or so before the court was to decide Dylan's fate. At the end of this time, I would have to submit a detailed evaluation of Dylan and his behavior to the court. His fate rested on us changing his behavior in this time.

That Dylan believed himself a leader was clear. As usual, I had to treat him holistically, removing that leadership from him using the whole repertoire of signals within the Amichien Bonding technique. In this particular case, however, I had to focus closely on moments of perceived danger. This had been when the attack had occurred. Only by teaching Dylan how to behave in this situation could I hope to save him.

It was not hard to see why Dylan had decided to be protective. Around the house, he and Helen were inseparable. His status was underlined by the fact that she allowed him to rush to the door, pull on the lead, and insist on cuddles whenever he wanted. When Helen started using Amichien Bonding, Dylan began to look at her in a completely different light, seeing that it was now Helen who was the decision maker and protector. It was no longer his function to look after the pack.

A week or so before the court case, I wrote my report. I did not believe Dylan was a threat any longer. My words to the judge were this: Dylan's owner realized she had been giving her dog the wrong signals, she now knew the correct signals, she would never allow the dog to be put in that sort of confrontational situation again. The magistrate was, of course, at liberty to ignore it. But it was my opinion that Dylan's behavior had been cured.

I always feel a sense of protectiveness toward the dogs I work with, sometimes too much I think. I must admit I lost the odd hour's sleep wondering how Helen and Dylan were going to fare. On the morning of the hearing, Helen rang me from the courtroom. She was close to tears and could only get out two words before breaking down. "He's safe," she said.

The magistrate had taken ten minutes to evaluate the case, then placed a controlling order on Dylan. It meant that she could keep him. Provided he did not attack anyone again, they could carry on their life together. I have now been involved in five such cases, and I am pleased to say that in each of them I have helped save the dog's life.

* * *

People have often labeled me a Pollyanna; they say I am too quick to see the good in others, to view every experience from the positive perspective as an opportunity to learn. I won't deny it—I do believe in treating life as a glass half full rather than a glass half empty. It was ironic then that, when my method proved itself in rather dramatic circumstances one day in 1998, I was the last to see it as a positive experience.

On a warm summer's evening, I had taken my pack of dogs out for a walk at a favorite beauty spot in the Lincolnshire countryside. I had loaded them into the car and headed for a footpath that ran alongside a pretty little stream. As we walked together, I vividly remember thinking what a wonderful evening it was. The sun was blazing low in the west, the birds were singing, and there was a lovely light breeze brushing itself against my face. The dogs weren't complaining either; they were freely running around, splashing in and out of the water. Life seemed pretty perfect in all honesty.

It was as we walked on that the idyll turned into a nightmare. The dogs had, as they often do, got ahead of me, which was perfectly all right as I knew they would all come back if I called them. For a brief moment, they disappeared from view around a right-handed bend in the path, when I heard a sudden scream. As I ran toward the sound, I almost tripped over Molly, one of the spaniels, who was rolling in front of me, crying and snapping frantically. As I looked ahead, I saw the rest of the dogs maniacally barking and jumping around as well. It took me only a second to realize what was wrong. Ahead of me was a row of bee hives. The dogs were being attacked by swarm after swarm of their inhabitants.

For the next few seconds, everything seemed to be happening in slow motion. As I struggled to gather my senses, I found myself under attack as well. It was one of the most terrifying experiences of my life. I cannot really explain the fear I was feeling. With the bees swarming all around my face, I could not see ahead of me. My ears were filled with the sounds of the buzzing around me and my dogs yelping and screaming in agony somewhere ahead of me.

I reacted instinctively and began moving as fast as I could toward the car, parked around 600 yards away. It was horribly slow progress. I tried waving my arms to little avail. I then started thrashing the air with my dogs' light rope leads which I had around my neck. To be honest, I was oblivious to the stings that were raining down on my head, neck, and hands. I just pressed on as best as I could, falling over regularly as I did so. Six hundred yards have never seemed so far.

Eventually I managed to get to the car. My hands were shaking so badly it seemed to take an eternity to get the key in the lock. I firstly opened the boot door and beckoned the dogs in. The dogs were all inside in what seemed like an instant. I then jumped into the driving seat, started up the engine, and opened the windows and sunroof so that the bees could escape. I then hit the accelerator as hard as I could and roared off. To my amazement, the bees outside stuck with us for more than a mile down the narrow lane. Eventually, however, we got to the open road and outran them.

I can't really remember the journey home. Back at the house, I got the dogs inside and began to assess the damage. Barmie had got off lightest, perhaps because he is so low to the ground. The spaniels, Molly and Spike Milligan, had been stung but only intermittently from what I could see. Their floppy, furry ears had protected their faces, although both had been stung quite badly on the lips. Ironically, it was the biggest and most powerful of my dogs, the shepherds, that had fared worst.

The worst was Chaser, the six-month-old son of Sadie. I saw that his right eye was closed completely. The swollen eyelid was a fiery red. When I called the vet, he agreed I should bring him to the surgery immediately. The other dogs were shaken but safe, so I felt able to leave them at home while I dealt with the worst victim.

At the surgery, we were treated by one of our regular vets, Simon. He took one look at Chaser and gave him an antihistamine injection, checking him again for any more stings. With the treatment over, I was able to relax for the first time in an hour. I think it was only then, as my adrenaline levels began to drop, that I began to be aware of the throbbing pain in my head and the various stings I had picked up on my face, neck, and hands. I must have looked a dreadful sight. I was feeling pretty sorry for myself; the experience had been one of the most traumatic I had ever encountered. Seeing my dogs in such distress was something I never wanted to repeat. It was only when Simon started asking me about the ordeal that I realized the significance of what had happened.

Simon knew me and my dogs well and asked me to explain what had happened. I ran through the story and he was horrified.

"How long did it take you to find all the dogs and gather them together again?" he asked me. "They must have been scattered for miles around." Only then did it dawn on me that in the midst of all that pain and chaos, my dogs had stuck by my side. I had not had time to even register the fact at the time. I had taken it for granted that they would be with me when I opened the door and so they were.

It was on the drive back to the house that the reality hit me. Despite the fact that they were faster runners, that they had the option of running in any direction they pleased, and that they were in extreme distress, my dogs had stuck with me. They had trusted me to get them safely out of the situation. They had proved my method worked in the most testing circumstances imaginable. That evening, back at home, I sat on the floor making an extra fuss of all my dogs at dinner time. I sat there for a while afterward, laughing as the tears ran down my face.

* * *

Perhaps the most satisfying aspect of my work has been the manner in which it has taken me in new and interesting directions in my life. In the autumn of 1998, for instance, I was asked to become a reporter for my local BBC radio station, BBC Humberside. I had been a regular guest on a phone-in program on the station for four years, answering callers' queries about dogs and their delinquent ways. The editors there seemed to be pleased with the response and invited me to do more work. The first piece was a diary of a day at Cruft's and proved popular enough for me to be asked to do a second piece. I must admit I was lost for words when they asked me whether I'd like to do an in-depth interview with none other than Monty Roberts.

By now the success of his book on his experiences, *The Man Who Listens to Horses,* had made Monty a worldwide celebrity. The success of the Robert Redford film *The Horse Whisperer* had added to the fascination with his unique, humane way of working with animals. It turned out Monty had returned to Britain and was staging a demonstration near the town of Market Rasen. He had agreed to talk to the radio station.

In the years since I first encountered him, I had seen Monty at work on around twenty horses. Each time, my respect for his work deepened. Each time, my certainty that man is capable of communicating with other species strengthened. I am not a trained journalist, so while part of me was thrilled at the prospect of seeing Monty at work again, part of me was petrified at the prospect of conducting an interview. I traveled down to Market Rasen with a blend of excitement and trepidation.

At the event, I met and spoke with Monty's regular associate in England, Kelly Marks. I was flattered in the extreme when Kelly, a former jockey who has become one of Monty's most trusted proteges, said she had heard of me and my work. I was utterly flabbergasted, however, when she then turned to Monty and said: "Hey, this is Jan Fennell." Monty was the same genial, unlikely looking cowboy I had first cast eyes on years earlier. He came over smiling warmly. "What's this I hear about you adapting my method for dogs?" he asked. "How've ya done that, lady?" I replied, "I listened to them!" and he laughed.

We chatted briefly before getting on with our radio interview, one of a few he was doing that day. To my delight, Monty then asked me to stay with him while he chose the horses he planned to use for his demonstration that night. It was all useful material for my radio piece, and I was delighted to accept. At the end of the afternoon, Monty asked me whether I planned to come back for the evening's demonstration. When I replied that I did, he asked me to come and say hello to him then. "Maybe we can do something together," he said as we parted company.

To be honest, I thought nothing more of his comment. I had enough to do making sure the interview was fine and getting home in time to see to my dogs, get changed, and get back to the event that evening. It was only when I got back to Market Rasen and saw Kelly again that it began to dawn on me that something was afoot. By now the stands were filling up. Such was Monty's drawing power, the 1,000 tickets had been sold out weeks earlier. Kelly asked me to join her in the middle of the arena next to Monty's around pen. I must admit I sought out the least conspicuous spot but, even then, felt terribly self-conscious.

Monty carried out his usual fascinating show. He went through two half-hour demonstrations, the first, "starting a baby," in which he saddled a horse never before ridden, the second in which he tackled a horse with a habit of kicking out at people. It was as the second half of the show began that I realized what Kelly and Monty had planned. As Monty walked back, Kelly ushered me into the famous around pen with her. When I hesitated for a second, Monty grinned and beckoned me in, coaxing me like some reluctant mustang he had just begun coaching. Before I knew it, Kelly was introducing me to the audience.

She made a brief speech in which she explained that Monty's method had been the inspiration for a number of other trainers. In the years since he had made his ideas public, he had been constantly surprised by the work these people had done. Kelly admitted that neither she nor Monty had been more surprised than when they heard of the work being done with dogs by an Englishwoman. I was, in truth, turning bright red with embarrassment at this point. Before I knew it, however, Kelly was bringing her speech to an end, telling the audience that I was about to explain my work and handing me the microphone. At first my heart was in my throat. Somehow I composed myself and began talking to the packed stands around me. I explained how seeing Monty had changed my life and how the remarkable results they had just witnessed with horses were also possible with dogs. It was only afterward, when people seemed to have understood what I had said, that I realized how fully formed my ideas had now become.

At the time, I must admit the whole thing was nothing more than a blur—apart, that is, from one image. As I passed the microphone back to Kelly, I heard the sound of the applause rippling around the arena. I turned around to see that it was being led by Monty himself. The journey I have traveled in the last nine years has been inspired by his work. His belief in man and animal working together in harmony underpins everything I have done. Now here he was accepting—and very publicly endorsing—my work. It was a hugely humbling moment in my life and one that I will never ever forget.

The 30-Day Training Guide

Introduction

The response to the publication of *The Dog Listener* has been truly humbling. I seem to have struck a chord with so many people who share my belief that humans and dogs can enjoy a peaceful and rewarding coexistence. The thought that many of these people are now successfully implementing the compassionate training method I outlined, and so are learning to understand the dog's own language, heartens me enormously.

The many words of support I have received since publication echo those I first heard on a much smaller scale when I began treating problematic dogs almost ten years ago. At the time, people raised on the idea of "obedience training," of sublimating the dog's will to that of its human "master," were overwhelmed to discover that dogs could coexist with humans while still exercising their own free will. They were amazed to discover that, as I like to put it, the best form of control is using the dog's self-control.

People's responses to the practicalities of applying my method have conformed to an equally familiar pattern. And it was here that the seeds of the idea for this new section were sown. From my earliest experiences in dealing with owners on a one-on-one basis, I quickly learned that no two dogs, no two homes, and no two relationships between dog and owner are the same. On the one hand, this is one of the main reasons why my work remains such a source of fascination to me. However, this also makes it natural that my work divides itself into two distinct phases: the home visit itself and the post-visit, or backup, service.

During the crucial first consultation, it is my job to explain and demonstrate the principles that underpin my work. From there, I go on to offer owners basic instructions on how to implement my method. Many owners are able to replicate my method almost instantaneously, yet just as many require ongoing supervision. This is no criticism of any of them; while, at heart, my method is a simple one, the practicalities of implementing it are at times demanding. Some people don't have the time or the resources to devote themselves as fully as others. Some have tried so many different methods, they cannot separate mine from the others. Equally, while I believe all dogs understand and react to my signals in the same way, every dog has a personality that can manifest itself in a different form of behavior. Some dogs are simply more strong-willed and resistant than others. As a result, I make sure I am always at the end of a telephone line to guide owners through the difficult moments.

The Dog Listener is in many ways the equivalent of the home visit. Within its pages, I outline my ideas and explain how they evolved through long and hard experience. It is, if you like, my mission statement. This additional section is the backup, the practical guide to putting that philosophy to work. Many of the people who have contacted me since reading *The Dog Listener* have asked me to develop specific ideas that I have only touched on. Most of them have wanted to apply my method to the practical realities of their own situations. Many people, for instance, own more than one dog. Others find it hard to impose themselves on their dog, to present the signals consistently, and to interpret their dog's reaction correctly. Some find it difficult to incorporate my method into their hectic lives. Others have dogs that have displayed behavior so bizarre, it simply hadn't occurred to me it existed! I have no doubt there are dogs out there with traits that I still haven't imagined. Nevertheless, I have done my best to cover as many of the unexplored areas as I can in the 30-Day Training Guide.

My goal is to allow anyone to implement my method. So, to make it as accessible as possible, I have divided what follows into a guide through the key phases of the crucial first thirty days. I am adamant in my belief that the work of a responsible and caring owner begins even before a dog arrives within the home. With this in mind, I have included a section on the all-important preparations required too. From meeting and corresponding with people, I know how many readers enjoyed hearing of my adventures in treating problem dogs, so I have made sure I have included as many anecdotal examples as possible to illustrate my points. In response to readers' comments, there are also step-by-step photographs.

There are goals that can be achieved at regular intervals throughout this method. For instance, it is my belief that most dogs and their owners can attain a basic new understanding of each other within two days. Within a week, I believe most owners should be ready to begin taking their dogs out on walks in public places. At the same time, I am not so foolish as to claim that my method will turn even the most difficult and badly behaved dog into the perfect companion within a month. Nor am I going to promise that every dog's progress will conform to precisely the same pattern. If you find your dog is not doing something within four, fourteen—or even forty—days, you should not despair. Results will come with application and patience. Provided you persevere and apply my ideas consistently, even the most desperate owner should see huge improvements within this first month. Perhaps even more importantly, you should also have changed your perspective of your dog. Both of you should then be ready to join the growing band of owners and dogs that are enjoying a happier and more fulfilling life together.

PART ONE:

Day 1

A Reintroduction to Dog Listening

It was around 12,000 BC that the modern dog, *Canis familiaris,* evolved from its ancient ancestor, *Canis lupus,* the wolf. In the centuries that have followed, the two animals have followed entirely different evolutionary paths. While the wolf has remained, to all intents and purposes, the same animal, the dog has multiplied into myriad breeds. While the wolf has remained in the wild, the dog has been domesticated. And while the wolf's life remains rooted within the same social environment—the pack—the modern dog has become integrated into human society, often living in isolation from other members of its species. On the surface, then, it may appear that the two have very little in common today. Nothing could be further from the truth.

At the beginning of the twenty-first century, our knowledge of our ancient *Homo sapiens* cousins deepens almost daily. Rather than fading with time, our understanding of their physiology and psychology seems to loom into focus more and more. And, as this happens, so science is growing increasingly certain that much of modern human behavior remains rooted in this Stone-Age past. Many of our most basic instincts, from our mating rituals to our attitude to other tribes, date back to the cave-dwelling experience of our ancestors. When you think about it, this makes perfect sense. In the great scheme of things, we have existed on this planet for barely any time at all. If the lifetime of the earth is a single day, man's time on the planet amounts to barely a few minutes. Our circumstances and surroundings may have changed rapidly, but our brains—and therefore our fundamental natures—have hardly evolved from those times. If this is the case with humans, then it makes sense to assume it applies equally—if not even more strongly—to our closest companions in the animal world, dogs.

When the wolf, *Canis lupus,* integrated itself into human society, the two species formed a unique partnership. Their relationship was so special that burial chambers have been unearthed containing the skeletons of men and dogs buried alongside each other. Both were hunter-gatherers, both were communal. Both understood instinctively that survival was dependent on the power of the pack. Since then, this new strain of wolf, *Canis familiaris,* has evolved into a multitude of breeds. Just as the roots of the entire European population can supposedly be traced to seven women, so every dog—from the Pekinese to the Saluki, the Akita to the Alaskan husky—has a bloodline that leads back to the first domesticated dogs. At the same time, the remarkably close, instinctive relationship ancient man had with the dog has deteriorated and all but disappeared.

If we accept that *Canis familiaris'* basic programming remains much the same as it was when it first left its wolf-pack environment, it is not difficult to understand the forces that mold a modern dog's behavior. The dog may have been taken out of the wolf pack, but the wolf pack can never be taken out of the dog.

To understand the way the modern dog views its world, we must therefore begin by looking at the society from which it first emerged and evolved: the wolf pack. For the wolf, the most powerful instincts are survival and reproduction. Driven by these instincts, the species has evolved a hierarchical system as strict and successful as any in the animal world. Every wolf pack is made up of leaders and subordinates, and at the head of every pack's pecking order are the ultimate rulers, the Alpha pair.

As the strongest, healthiest, most intelligent, and most experienced members of the pack, it is the Alpha pair's job to ensure the pack's survival. As a result, they dominate and dictate everything the pack does, and their status is maintained by consistent displays of authority. Underlining this, the Alpha pair are the only members of the pack who breed, thus ensuring only the healthiest genes survive. They are, in effect, twin dictators. They control and direct life within the pack, and the remainder of the pack accepts that rule unfailingly. Each subordinate member is content to know its place and function within this pecking order. Each lives happily in the knowledge that it has a vital role to play in the overall well-being of the pack.

The hierarchy of the pack is constantly reinforced through the use of highly ritualized behavior. The ever-changing nature of pack life, in which Alphas and their subordinates are frequently killed or replaced through age, makes this essential. As far as the wolf's modern-day descendants are concerned, however, four main rituals hold the key to the pack instinct that lives on within them. They are central to my method.

The Four Rituals

1 The first key ritual is performed whenever a pack is reunited after being apart. As the pack reassembles, the Alpha pair remove any confusion by reasserting their dominance via clear signals to the rest of the pack. The pair

have their own personal space, a comfort zone, within which they operate. No other wolf is allowed to encroach on this space unless invited to do so. By rejecting or accepting the attention of other members who wish to enter their space, the Alpha pair re-establish their primacy in the pack—without ever resorting to cruelty or violence.

2 When a kill has been made by the pack, the Alpha pair get absolute precedence when it comes to eating the carrion; the pack's survival depends on their remaining in peak physical condition. Only when they are satisfied and signal their feed is over will the rest of the pack be permitted to eat—and then according to the strict pecking order, with the senior subordinates feasting first and the juniors last. Back at the camp, the pups and "babysitters" will be fed by the hunters regurgitating their food. The order is absolute and unbreakable. A wolf will act aggressively toward any animal that attempts to eat before it. Despite the fact that the pack contains its blood relatives, an Alpha will attack any wolf that breaks with protocol and dares to jump the queue.

3 The Alpha pair repay the respect the pack bestows upon them with total responsibility for its welfare. Whenever danger threatens, it is the role of the Alpha pair to protect the pack. This is the third situation in which the natural order of the pack is underlined. The Alpha pair perform their leadership role unblinkingly, and from the front. They will react to danger in one of three ways, selecting one of the "three Fs": flight, freeze, or fight. Accordingly they will run away, ignore the threat, or defend themselves. Whichever response the Alpha pair select, the pack will again back up their leaders to the hilt.

4 Inevitably, the Alpha pair are at their most dominant during the hunt. Food, after all, represents the pack's most fundamental need; its very survival depends on it. As the strongest, most experienced and intelligent members of the pack, the Alpha pair take the lead during the search for new hunting grounds. When prey is spotted, they lead the chase and direct the kill. The Alphas' status as the pack's key decision makers is never more in evidence than during this process. The wolf's prey can range from mice to buffalo. A pack may spend as long as four hours stalking, cornering, and slaying its target. The logistics of this operation require a combination of organization, determination, tactics, and management skill. It is the Alpha's job to provide this leadership. It is the job of the subordinates to follow and provide support.

A Leap of Imagination

It is little wonder that so many people encounter problems in their relationship with their dogs. The vast majority of dog owners enter that relationship having made a series of utterly false assumptions. They have assumed, for instance, that the dog is in effect a child: an incapable, illogical—if deeply lovable—dependant. As a result of this, they have also assumed that it is going to respond to a series of childlike instructions issued in their language. They are viewing the dog and its world through the prism of the human experience: they are wrong.

The dog does not understand or relate to the human experience in any way. The fact that it responds to the word "sit" or "come" after hearing it a thousand times does not mean that it understands the spoken human language. It has simply come to associate these sounds with certain forms of behavior, and has learned to act accordingly.

To lead a happy and fulfilling life with dogs, all owners need to make a fundamental change in their approach. They need to look at the world from the dog's perspective, and understand the society and the rules under which the dog believes it is living. A leap of the imagination is required. They need to understand that the dog is not an immature child operating within the human world; it is an intelligent adult operating within the structures and strictures of its own highly regulated society: the pack. And, most importantly of all, the dog believes it has been given the job of leading that pack. Viewed from this perspective, all its behavior will make sense. And, viewed from this perspective, all owners have the opportunity to forge a new and hugely fulfilling relationship with their dog.

Amichien Bonding

A dog believes it is a functioning member of a community that operates according to principles directly descended from the wolf pack. Whether its "pack" consists of itself and its owner, or a large family of humans and other animals, the dog believes it is part of a social grouping and a pecking order that must be adhered to at all times. What is more, all dogs believe they are the elected leader of that pack. And they believe this because the humans with whom they live send out signals that re-elect them to the role on a daily basis.

It is my belief that all of the problems we encounter with our dogs are rooted in their belief that they rather than us, their owners, are the leaders of their particular packs. It is the correction of this misconception that lies at the heart of the communications technique I have developed, called Amichien Bonding.

Four separate elements make up the bonding process. Each correlates to the specific times I have identified when the pack's hierarchy is established and underlined. On each occasion, the dog is confronted with a question that we must answer on its behalf by providing clear, concise signals that it will understand. The four occasions are:

- When the pack reunites after a separation, who is the boss now?
- When the pack eats food, what order do we eat in?
- When the pack is under attack or there is a fear of danger, who is going to protect us?
- When the pack goes on the hunt, who is going to lead us?

Each set of signals will be introduced individually with the "hunt," or walk, tackled last of all. It will eventually be necessary for you to use all four elements in conjunction with each other. Your dog must, in effect, be blitzed with signals. It needs to learn that it is not its responsibility to look after you, that it is not its job to care for the house, that all it has to do is sit back and lead a comfortable and enjoyable life. It is a mantra that must be repeated over and over again. Only then will your dog get the message that it is no longer in charge; only then will it be able to exercise the most powerful form of control, self-control.

The process is, at heart, a simple one. The most powerful central principles are established within the first few hours of application. From there, you will go on to add additional signals and controls, first within the home, then within the outside world. By the end of the first thirty days, you should have laid the foundations for a lifetime of companionship and co-operation. What follows is a guide to these formative, first thirty days.

In many ways, the process that is about to begin is akin to a revolution. The objective will be to stage a bloodless coup, to remove the leadership from the dog, but to do so without resorting to force or violence in any way. No coup in history has succeeded without precise and careful planning. And this one will be no exception . . .

To Be or Not to Be a Dog Owner

"THERE IS NO FAITH WHICH HAS NEVER YET BEEN BROKEN, EXCEPT THAT OF A TRULY FAITHFUL DOG."
Konrad Lorenz

It is no surprise to me that dogs have for centuries remained the inspiration for words like these. The love, loyalty, and companionship that a dog can bring into our lives are unique. And, in theory, they are pleasures that everyone should have the right to enjoy. I write "in theory" because I do not, in all honesty, believe it is a right that should be granted automatically. It is a sad fact of life that there are people who take dogs into their home for all the wrong reasons. The animal sanctuaries and dogs' homes of the world are filled with the unhappy results of their rash and unthinking behavior.

It has become a cliché to say that a dog is not for Christmas, it is for life. But it remains true nevertheless. A dog brings with it a set of responsibilities. And no one should take on the task of introducing a dog into their home without weighing up whether they are ready, willing, and able to live up to these responsibilities.

There have probably been thousands of books written on the pros and cons of taking on a dog. I do not intend to contribute another one. Having said that, I do have some opinions on the matter. Anyone who is only getting a dog for their children to play with should not get a dog. Dogs and young people can forge the most beautiful relationships of all—but only if the child is taught to respect the dog. They are not toys. Anyone who is getting a dog solely for the purpose of guarding their homes should not get a dog. This is not fair. Anyone who intends getting a dog and then leaving it at home all day should think again, or should make appropriate arrangements for a friend or professional walker to take it out. The dog is a social animal and, if it is cut off from other beings for eight to twelve hours, this is not good for it.

I also believe that all owners must prepare themselves for the reality of dog ownership. It is the easiest job in the world when the dog is a lovable ball of fluff and affection, but what about when it falls ill or misbehaves, fouls the living-room carpet or growls at visitors? And what about heading out into the outside world? Are you prepared to head out on cold, inhospitable winter mornings to walk the dog? Are you willing to become a fully paid-up member of the plastic-bag brigade and clean up after your dog in public? And what about the time it falls ill and you face the potentially astronomical vet's bills?

Of course, good owners—rather like good parents—think about a lot of these things instinctively. But I would ask even these genuine dog lovers to do a little soul-searching before committing themselves to using my method. It is one thing to understand the challenges of my approach from the human point of view, but what about thinking about this from the dog's perspective? Dogs are living, breathing creatures with needs and feelings of their own. No one would welcome a deaf person into their family without accepting in advance that they would need to learn some form of sign language. It is no different with a dog. So, for this reason, I would ask you to consider the following:

- Are you prepared to go into this with an open mind?
- Are you willing to forget and discard all other "training" methods?
- Are you prepared to accept the dog's welfare is paramount?
- Are you prepared to work hard and make the sacrifices necessary to understand a language that may prove elusive and confusing at times?

- Are you prepared to give the time the dog needs, especially during the early stages when patience and perseverance are of primary importance?
- Has the whole family discussed what is going to be required?
- If you already own dogs, are you prepared to go through the necessary retraining they will need as well?

These are all tough questions. I make no apologies for posing them, however. I much prefer that people understand their dogs before they learn to love them. I would rather they begin knowing that sacrifices and hard work will be required. Yet, if my years of experience have taught me anything, it is that the effort is far, far outweighed by the rewards that come from having a happy, well-adjusted, and relaxed dog.

Where to Get a Dog

To my mind, there are only two sources from which we should buy dogs: respectable, responsible breeders and registered rescue centers or dogs' homes. There is, I know, a wide range of alternative sources, from pet shops to so-called licensed kennels. Yet I would not recommend that a dog be taken from any of these, for the following two reasons.

Firstly, it is my firm belief that, if at all possible, an owner should be able to see the dog's mother and gain knowledge of the history of both parents before taking the dog away. This is something I will explain in more detail in a moment. Secondly, it is only through the first two sources that an owner will have any comeback.

As an illustration of this latter point, I often use the example of a close friend of mine, Wendy Broughton. Wendy is a keen horsewoman. It was Wendy who introduced me to Monty Roberts, the Horse Whisperer, from whom I learned so much. At a horse fair one day, Wendy saw a batch of dogs for sale for $25 each. There were no clues as to where these dogs had come from. Their owners that day had only one interest: making money. Wendy is a soft touch and took pity on one dog in particular. In the two years since then, she has spent more than $4,000 on vet's bills. It turned out the dog was riddled with worms and suffered from a variety of stomach problems.

The crucial point here is that Wendy had no comeback on the people who sold her that dog; she didn't even get a contact name. This type of trade is sadly on the increase. The growth of "puppy farms" is something I abhor. Often, owners will be allowed no more than seventy-two hours in which to return their dogs.

In contrast, no reputable breeder or rescue center would sell a dog without guaranteeing that you could return the animal to them if there were a medical problem or if you were, for whatever reason, unhappy with it. Indeed, a good breeder would insist on being on hand to provide advice throughout the dog's days. The dog cannot lose under this arrangement. If it fits into a happy home, it will enjoy a long and hopefully rewarding life there. If not, it will return to a place where its best interests are safeguarded. Owners who leave themselves with no comeback are left with no obvious place to return an unwanted dog. Many will, of course, ensure they are given good homes. Sadly, many more will not.

Buying a Puppy

Falling in love with a puppy is the easiest thing in the world. We've all experienced the "aaah" factor, that moment when we've stared for the first time into the saucer-sized eyes of a sweet, young dog and gone hopelessly gooey. It's for this reason that I recommend you begin your search for a puppy by steering clear of these charmers to begin with. You know that you are going to fall head over heels in love with the puppy—it's a given. Far better for you to deal with the realities before beginning the romance, and the best way to do this is by first meeting the family.

Producing a litter of puppies involves collaboration between the human breeder and their dogs. The dogs deliver this new life into the world, but it is the human who must shoulder the ultimate responsibility. I passionately believe that the best way to assess whether a dog is coming from a good home is by meeting both sets of "parents," that is both the human and canine carers.

Given that many breeders go to a stud to begin their litters, it is more than likely that only the mother will be available. The condition of the owner and its home, personality, temperament, and general demeanor will reveal much. Of course, part of the equation is missing if the father is not there. Seeing both will provide the clearest idea possible of the sort of dog their offspring is going to become. Nevertheless, a visit to the mother will always be worthwhile.

Just as importantly, the attitude and behavior of the owner is highly revealing too. If, for instance, the mother is not with its puppies, there are immediate reasons for concern. The growth of puppy farms is one of the more unpleasant facts of life these days. The absence of the mother may mean that the seller is a third party, whose motives are purely to do with money rather than the welfare of the dogs in their protection.

Another good indicator of the quality of the home is the extent to which the seller interrogates all potential new owners. A potential buyer should expect to be grilled by a good breeder. I know, as I always insist on finding out as much as I can about anyone who wants to take one of my puppies into their home. The procedure should be like that of an adoption society being very careful about where it places the children entrusted to its care. Good breeders should be just as diligent in finding a home for the innocent young lives for which they are responsible.

A reputable breeder should want to know everything about the home into which they are considering releasing the puppy. They should want to know what the domestic situation is at home: are there people around all day to look after the dog? Are there many small children in the home? Has the prospective owner owned this type of dog, or indeed any dog, before? Another important question is whether the prospective owner has thought about whether a particular breed is right for their home. Also, are they willing to wait for a puppy? Good breeders do not churn out dogs as if from a production line. If the breeder asks these questions, it should be seen as a positive sign. If they do not, they are probably more interested in making a sale and should therefore be treated with caution.

By the same token, a good breeder should be open to questions from a potential owner. They should be willing to reveal anything and everything about the dog's history and background, from the details of its parentage and its age, to its favorite food and toys. Again, owners should be wary of anyone who is vague or unhappy about answering these questions. I've never been afraid to talk to potential owners in this kind of detail. Indeed, I've made very good friends of people who have bought puppies from me.

Potential owners should be particularly careful about checking for hereditary problems within some breeds. Cavalier King Charles spaniels can be prone to heart problems, for instance; Dalmatians can suffer from deafness; long-backed breeds, like basset hounds and dachshunds, have a tendency to suffer from back pain and slipped discs. With German shepherds and Labradors, potential buyers should look out for hip dysplasia, a genetic problem in which the ball-and-socket joint of the hip can be deformed or even nonexistent. It is a condition that is extremely painful and ultimately crippling for dogs that are afflicted, and something that good breeders monitor closely. Dogs are x-rayed at the age of one and given a "hip score," which ranges from zero for perfect hips, to eighteen and higher, a mark that indicates the dog should not be used for breeding.

All good breeders work hard to eliminate these problems. But, so as to make informed and appropriate choices, potential owners are well advised to thoroughly research the breeds they are interested in. This is easily done through the Kennel Club or its equivalent organization, and then through the various breed experts. The importance of this cannot be overstressed. No one goes out to make a major purchase like a new car or a house without checking out the subject thoroughly. When they choose a dog, owners are introducing into the family a new member that will hopefully remain with them for twelve to fourteen years; it is not something to be taken lightly.

Rescue Dogs

Few things can compare with the pleasure that comes from providing a home for a rescue dog—that is, a dog that has been placed in a sanctuary or dogs' home due to abuse, abandonment, or bad behavior. As someone who has taken in a number of rescue dogs over the years, I can say with my hand on my heart that the rewards both the dogs and I have had have been phenomenal. The joy I have had in seeing tragic dogs given a new hope in life has been immense. And I have to admit it has made me feel good to have provided these dogs with a happy and stable home life they had previously been denied.

Having said that, there is no escaping the fact that rescue dogs can present considerable problems, for self-evident reasons. I often say that there is no such thing as a problem dog, but there are dogs with problems. And rescue dogs, by their very nature, come with more problems than most.

Obviously in such cases it is simply not possible to make the checks that are possible with a new puppy. Any decent sanctuary or rescue organization should be able to provide you with some details about the dog's immediate past. Again, vagueness or reluctance to give out details should be regarded with extreme caution. However, the vast majority of sanctuaries and rescue homes are run by people who have an extraordinary dedication to dogs and their welfare. Whatever they know about the dog, they will be willing to tell you—good or bad. It is not in their interest, or that of the dog, to deceive you. The reason many dogs are in a sanctuary is due to violence. I believe, however, that, provided you understand the way a dog's mind works, it is a calculated risk. For many people, I hope it will remain a risk worth taking.

A Question of Breeding

When I was given my first dog, a Border collie called Shane, there were relatively few breeds readily available to buy. How things have changed. Today, new owners face a bewildering choice of breeds, from exotic Oriental dogs like the Akita and the shih-tzu, to breeds that, until recently, were considered rare, like the Newfoundland or husky.

I am often asked whether particular breeds are more suitable for some people than others. My answer, in general, is that anyone can own any breed provided the dog fully understands its status within the domestic pack. A Yorkshire terrier that believes it is boss may not deliver as bad a bite as a Rottweiler or German shepherd under the same delusion; nonetheless, a bite is a bite. If a dog's attitude is right, this problem will not arise, so the breed of dog is immaterial; yet there are some general guidelines you should bear in mind.

If, for instance, you have a quiet, family lifestyle and are looking for a more relaxed companion rather than an active dog, you might be advised to take a breed that does not need as much exercise, something like a Tibetan spaniel. This does not necessarily mean a smaller dog: Great Danes and Saint Bernards love to lie around. On the other hand, people, like me, who find it hard to justify going out for a walk without a dog for company, are spoilt for choice. German shepherds, Labradors, springer spaniels are all suitable companions.

You have to bear in mind the specifics of your situation. For example, I would not advise someone who is restricted in what they are able to do physically to take on dogs with heavy coats, breeds like the Afghan hound or rough collies. The demands these breeds make in terms of grooming are considerable. It is far better for people in this situation to go for a Labrador, a retriever, or a Border collie, dogs that are less "high maintenance" in this respect.

You should also bear in mind the practicalities of training a dog. For instance, with the introduction of basic controls such as the "sit" and "heel," you will be required to do a lot of crouching and bending down to deal with your dog. Older, less mobile, people might want to bear in mind whether they are going to be capable of doing this with a small dog in particular. It is a lot easier to reach the eye level of a German shepherd than that of a lhasa apso.

It is worth stating at this point that it is wrong to associate the size of the dog with the size of the task ahead of its new owner. Big dogs do not necessarily mean a bigger task in terms of training and maintaining them. Indeed, in my experience, some of the kindest, gentlest, and easiest-trained dogs have been from the larger breeds. And some of the most uncooperative have been from the smaller ones.

Of course, the best way to understand the type of dog you are acquiring is to study the breed. Each breed was originally evolved, through human intervention, to perform certain functions. Border collies and German shepherds were bred as sheepdogs, and are carers and herders. Beagles were bred to hunt. The Saint Bernard and the Newfoundland were bred as rescue dogs—it is why, in normal circumstances, they are such peaceful, placid dogs. It is only when an emergency arises that they leap into action. By the same token, lhasa apsos and Pekinese were bred as lapdogs. Breeds like this are, literally, custom-made for people who are looking for warmth, affection, and companionship above all else. There are dogs to suit everyone.

The one thing I would caution against is getting a dog purely because it is in vogue at a particular time. Certain breeds come into fashion every now and again. I can recall times here in Britain when the Afghan hound and the Old English sheepdog became incredibly "trendy" dogs. The success of the film *101 Dalmatians* suddenly made that breed popular. A short time later, when these breeds are no longer *de rigueur*, the sanctuaries and rescue centers are full of unwanted dogs. In the meantime, unscrupulous breeders will have gone into overtime producing as many puppies as possible, inflicting God-knows-what damage on the breed itself. If someone is going to choose an accessory from a fashion magazine, let it be a handbag, a pair of shoes, or a dress—anything but a dog.

Expanding Packs: Preparing for a New Dog

The more I have observed and worked with humans and dogs, the more I have come to recognize our shared sense of family values. It is, of course, no accident that dogs possess a pack instinct as deep-rooted as our own. It is one of the reasons why our ancient ancestors forged an alliance with the dog, the first and most important animal to be domesticated by man. I mention it at this point because you must consider this when you expand your pack by introducing new dogs to a home that already has dogs. My method, as I have explained, allows you to position yourself as the leader of that pack. Yet, in situations where you already own two or more dogs, the dogs, too, will have a hierarchy. The natural result when a new dog is introduced into the pack is a situation where the dogs believe a leadership election is about to happen. I will deal in detail with this later on, but for now bear in mind that the introduction of a new dog is something that has to be planned with particular care.

There are many other factors to consider. Mixing breeds, for instance, has to be thought about. A Japanese Akita and a German shepherd will get along together eventually, but there may be a great deal of friction before they settle into a routine together. Introducing a puppy less than nine months old to an existing pack is, in relative terms, a straightforward process. But here, too, you must bear in mind that placing a nine-month-old puppy in a home is the equivalent, in dog years, of introducing a hyperactive twelve-year-old to a human household. People always talk of one dog year being the equivalent of seven human years; in the case of the dog's first year, however, this is more like sixteen years. Look at it this way: by nine months old, a bitch is physically capable of producing and raising children. Emotionally, however, she is nowhere near ready. You should always bear in mind the effect a newcomer like this might have on the chemistry of your existing pack.

If you already have dogs, I therefore suggest you do two things before expanding the pack. Firstly, I recommend you introduce my method to the existing dogs first. This makes obvious sense on every level: the influence the existing dogs have on their new companion will be immense. If they have learned to live a fulfilling life by accepting you as their leader, the chances of the potentially disruptive newcomer accepting the status quo are hugely increased.

Secondly, I also advise you to introduce the new dog to its prospective pack as soon as possible. The ideal option is to take the existing dogs to the breeder or rescue center where the new dog is. This has several obvious advantages. In territorial terms, it is neutral ground. The new arrival will also be among people who know and can control it in the event of problems. Most importantly of all, if after a few meetings there is clearly a high degree of friction between the old and new dogs, you will have the chance to think again. Once more, you will be able to get a strong idea of whether a refuge is a reputable one or not from their handling of this whole process. Alarm bells should begin ringing if they are reluctant to allow you to bring your dogs to meet the potential family member. The danger signals become overwhelming if the refuge even hints at the sale being cut and dried.

Preparing the Home

A dog does not really care whether its owner lives in a tent or a palace. The criteria it regards as important are more to do with the comfort, safety, and quality of life it enjoys rather than the particulars of its surroundings. If only we humans were less materialistic sometimes—as long as a dog is cared for in the right way, it will be happy.

In preparing to put my method into practice, I would advise you to make a few specific provisions around the home. Some of them are obvious, others less so. Each is important if life with a new dog is to get off on the right footing.

Sleeping Quarters

Decide where the dog is going to sleep. It is vital that the dog or puppy has its own space, its own refuge. It must be able to lie down on its side with about three to four inches to spare either side. It should have about the same space when it stretches out lengthwise. I personally like to use dog cages; my dogs seem to enjoy the idea that they have their own secure, enclosed space. I always leave the gate open so they can move freely. They are also perfect for traveling in the car.

A dog does not need to have an expensive basket or kennel of its own; a supermarket cardboard box turned on its side provides a safe, secure space. I always find putting an old jumper in it helps. It provides warmth and cover but has the added advantage over a clean blanket that it also smells of the family.

No-Go Areas

Decide on the no-go areas in the home. There will be areas where you will not want the dog to stray, typically bathrooms and toilets, maybe a study or dining room. You should establish these areas in your mind in advance. Every member of the family should also be aware of the rules; instilling them into the dog will prove that much easier later on.

Gates: With the above in mind, I recommend placing a children's gate between the kitchen and the hallway. It ensures that the dog does not follow you everywhere, something that is a particular trait with puppies. At the same time, you will be able to keep an eye on the dog's movements. The only thing to watch out for is that the gaps between the bars are not wide enough for the dog to get its head stuck.

Garden: The garden area will be of vital importance to both you and your dog, not just during the first thirty days but also throughout your lives together. You should ensure that the area is secure and that your dog cannot escape. As with the internal gates, you should also make sure there is no danger of the dog's head being trapped in fencing. In assessing this, I do warn owners that I have had to extricate puppies from the most unlikely places; don't underestimate their ability to stick their heads in the tightest spaces!

Basic Needs

It may be stating the blindingly obvious here, but there are still many owners who seem to overlook the most basic practical preparations for the arrival of a new dog. The list that follows is customized to allow you to apply my method immediately.

Feeding Bowls: Two bowls are essential. One is for water, which should be filled with fresh water twenty-four hours a day and should be changed at least two to three times a day. The other is for food. As you will see later, an extra supply of bowls—up to four per dog—may come in useful in applying part of my method.

Collar and Lead: It will be several days before the dog is ready to go on a walk. Before then, however, there will be times when it will be used for heel work inside the house.

Toys: Again, there is no need to go to any extraordinary expense here. For puppies in particular, I find there is nothing better than an old tea towel or blanket, soaked in water, then tied in a knot. It is perfect for young dogs with immature, malleable teeth.

Grooming Brushes: Grooming is an important element of looking after a dog. As well as keeping the dog in condition, it allows you to keep a regular monitor on your animal's weight, skin condition, and general welfare. It is also another means of forging a close personal bond with a dog. Long-haired dogs like cocker spaniels and Old English sheepdogs cannot be left for more than six days without being brushed. The knotted coat that results after this time can be painful and extremely unhygienic. Owners will need a "slicker" brush for these dogs. For smooth-coated dogs, a soft brush will suffice. I also find it useful to use a pair of rubber gloves. They can be perfect for scrubbing particularly tricky areas clean.

First-Aid Kit: In case of emergencies, it is advisable to have a collection of medicines, such things as wound dressings, ointments, and ear drops.

Food Rewards: An essential element of the work that lies ahead involves the awarding of tidbits of food. Before your dog arrives, ensure there is a good supply of special treats, things such as reward sticks, meat strips, or (my personal favorite) pieces of cheese.

Good, Local Vet: I suggest to all potential dog owners that they spend a little time checking out the vets in their area. It is a good idea to get the dog registered in advance.

Mental Preparation

By far the most important preparation is that which must go on in your head. Unless you have grasped the central principles of my method, you should not proceed with it. You must be ready to see every situation in the right context—in terms of the dog and its view of its place in the domestic "pack." In addition, you must also make practical preparations for the days ahead.

Never Set Yourself Up for Failure

One important piece of advice I like to get across to owners from the start is that they should never set themselves up for failure. By this I mean that, in the days ahead, there will be some setbacks that you cannot predict—and they are probably going to be related to the particular traits of your dog's personality. But, at the same time, there are going to be some mistakes that you can avoid through proper, careful planning—for instance, you should always have a ready supply of food reward with you. Try to be prepared for every eventuality that is thrown at you.

Allow More Time

Absolutely central to everything I do is a sense of calm. For this reason, you must be prepared to allow more time for everything; if you are impatient, it won't happen. You must be ready to put your dog's needs high on your agenda for a month. You must be ready to put yourself out for your dog. And, throughout this time, you must remember that you are taking into your home a creature whose language and needs are different from yours. Most crucial of all, you must accept that the dog is not going to learn your language; it is up to you to learn the dog's language. And, for this reason, every grown-up member of the family must take some time to familiarize themselves with the principles of bonding before the process begins.

Think Canine

A good owner should always anticipate their dog's doubts and questions. You should try to think ahead and recognize the things that a dog is going to find confusing or frightening. For instance, a dog is not being stupid if it recoils when a vacuum cleaner or a dishwasher or washing machine is switched on. I ask people to imagine how they would feel if they didn't comprehend what they were encountering. They'd want to know whether this was something that was going to cause them harm.

Be Ready, Willing, and Mentally Able

Learning my method is akin to learning a new language. Anyone who has had to master this skill will know what a difficult and time-consuming business it can be. They will also know that it requires absolute concentration and commitment, and a mind that is receptive and open to new ideas. It is no different with my method: you must be not only willing and able to learn, but also mentally ready. This is why I do not recommend you begin at times of emotional stress. Far better to postpone the introduction of Amichien Bonding until your mind is clear and focused.

PART TWO:

The First 48 Hours

New Horizons: Settling a New Dog in Its Environment

The road ahead is a long, sometimes difficult, but hopefully richly rewarding one. And for some owners, the first potential wrong turn looms within the first moments of bringing a dog home. Dogs that are already acclimatized and settled into a home will be able to begin the first phase of my method—reuniting after separation—immediately. For dogs that are new arrivals, whether they be puppies, adults, or rescue dogs, there is a preliminary hurdle to be overcome. It is vital if the new relationship is to get off to the correct start.

Most of us can vividly recall moments like our first day at school, or our arrival at our first workplace. It is an extremely stressful, daunting, and, at times, downright terrifying experience. People are operating according to a set of rules we don't yet understand. The place is filled with unfamiliar sights, sounds, and smells. Imagine someone having to fit into a new environment like this when nobody understands what they are saying. This is the situation a dog faces when it takes its first tentative steps into a new home.

Arriving in a new environment is a traumatic experience for a dog. The most important thing you can do during the first forty-eight hours or so is to make this as trauma-free as possible. And it is imperative that the first minutes of the day are spent allowing the dog to settle into its new environment.

The first thing I recommend you do with a new dog or puppy is that, on arrival at the new home, you immediately take it outside to do its toiletting. When the dog relieves itself as you have asked it to, you should reward it with a tidbit. This should be accompanied with warm words of encouragement, such as "Good dog," or "Clean dog," and perhaps a stroke of the head or nape of the neck area. The key point here is that the first piece of positive association has been achieved within the first few moments in the home. A good start has been made.

The next stage is to allow the dog to get to know its new environment. You should leave it alone and let it explore the areas it is free to roam in. Throughout this time, you should be giving out gently affectionate signs, by which I mean smiles and words of reassurance and kindness. At the same time, you should not be gushing or overaffectionate with the dog. This is particularly important in the case of rescue dogs. These are dogs that have had appalling experiences in the outside world. For understandable reasons, their trust may be much harder to win. The key thing here is to remain kind and calm but to keep a respectful distance too. It may well be that the dog is too nervous to engage in any way during the first few hours, or even days. Do not rush the dog; give it space and time.

* * *

Settling a dog down into a new environment may sound straightforward. It is not, I can assure you. As an illustration of the hard work that is sometimes required here, I often draw on the example of Murphy. All dogs need to check out their new home, to sniff out the landscape, if you like. Murphy was more qualified than most for this job. He was a police "sniffer" dog. At work, his role was to locate drugs, explosives, and contraband.

I met him while visiting his handler, Teresa, a PC with the West Midlands Police. She had expressed an interest in learning more about my method and I had visited her at her home. Murphy, I discovered, lived in a kennel at the rear of the house. When I wondered why he didn't live in the house, she simply laughed. "He'd destroy the place," she said. By this, she meant that Murphy was simply too inquisitive. His training had encouraged him to stick his nose in anywhere and into anything suspicious. On the rare occasions he had been allowed in people's homes, this had led to disaster. He would knock things over and generally leave a trail of destruction behind him. In short, he had no house manners.

I offered to illustrate my work by introducing Murphy to the house. He immediately went berserk, leaping around everywhere. I immediately asked for his lead. While Teresa and I sat down and talked, we studiously ignored Murphy's high-octane behavior. Throughout, he remained attached to his lead, which Teresa clung to from her chair. After ten minutes or so, he settled down. At this point, Teresa released him from his lead. The reprieve was short-lived; the moment he began leaping around again, I got Teresa to put him back on his lead. She didn't scold him or shout at him. She simply remained aloof and acted quickly. This process was repeated four times in all.

The next time Murphy was let off the lead, he simply sniffed around our legs. It was as if he was making his final check of the environment. Soon after that, he lay nicely down at Teresa's side. It was the final symbolic gesture, the signal that he was now ready to begin living, and learning, there.

* * *

It is no surprise that many people have problems integrating new dogs into a home that already has one or more dogs living there. My method will ensure that the human members are elevated to the senior positions within the extended pack, but it is important that you acknowledge the canine pecking order below this. This hierarchy is the most natural thing in the world, and nothing is going to change that.

In many, many cases dogs settle into a natural order with very little trouble indeed. But, make no mistake about it, the process can be hugely competitive and can result in aggressive and potentially violent behavior. For this reason, I must emphasize how essential it is that the initial, introductory meeting between dogs happens on neutral territory. In an ideal world, I would recommend more than one meeting.

If, for whatever reason, your existing pack is only going to meet the new dog on the day of its arrival, this introduction must be your first task of the day. You must choose a neutral ground, and here you must be careful to avoid places where the existing pack has strong associations. In the past, I have taken new additions to my pack to parks or open spaces, areas where they and my existing dogs have been free to weigh each other up and begin the process of integration in a less charged atmosphere. It requires the assistance of someone else to lead either the existing pack or the newcomer to the spot. Even if this only lasts for half an hour before the new dog arrives home, it can be of huge long-term benefit.

If there is more than one dog in the existing pack, I suggest you introduce these dogs to the newcomer one at a time. Each time this happens, it is vital the two dogs meet on equal status. As the newcomer is going to be on a lead, the existing pack member should also be on a lead. If one dog is being rewarded with toys or treats, then so should the other. The vital thing here is that the two dogs form a new, extended pack. By exuding strong leadership, you should be implanting the idea that the job of Alpha within this pack is already taken. The newcomer will obviously be less well versed in your communication methods than the established dog. Yet the message should begin to seep through just the same: by all means sort out a pecking order among yourselves, but be certain that the job of leader is not up for grabs—nor will it ever be.

As you allow the dogs to get to know each other, the key message here is "don't panic." If the dogs grumble or growl at each other, ignore it until they relax. Ultimately, dogs enjoy each other's company. Given time and space, a friendship can emerge from the least likely pairing. Throughout this period, it is again important to show the qualities of leadership at all times. So everything should remain calm, and you should keep your head.

It is worth stating that you should adopt a sensible approach toward aggression between dogs here. I have seen people overreact completely to the sight of two dogs squaring up to each other and growling. In human terms, this is no more serious than one person telling another to get out of their sight. It is not the most pleasant situation, but neither is it an excuse for the sort of histrionics I have witnessed in my time. If your dogs do threaten each other, you should act decisively and immediately to remove the instigator. For this reason, I would recommend that the new dog wears a collar while it is being introduced to the pack; it is better to be safe than sorry.

Once a rapport is established, let the dogs move closer together. Eventually, they should come up close enough so that they can play together. This is the most natural thing in the world. Again, remain relaxed and let it happen naturally. The more time the new extended pack spends together in this way, the better. Eventually, however, they will have to head home. The dogs should be transported home together. If the established dog is being reluctant getting into a vehicle, you should put the new dog in first. If the situation becomes tense or confrontational during the journey, the new dog should be moved to another seat. Once at home, the dogs must remain on equal status, each of them on a lead. They should be released in the garden at the same time and, while you keep an eagle eye on them, they should then be left to get to know each other.

There is every chance, of course, that the dogs will not get on immediately. You will have to exercise common sense and flexibility here. I strongly recommend that you should make plans for segregating your dogs at home in the early days. Dogs are quick to assimilate themselves into environments; in most instances, I believe dogs can settle into a new home within forty-eight hours. Yet, given the potentially explosive situation that can arise when a dog is added to an established pack, I ask people to be overcautious. I have certainly found that providing both the newcomer and existing dogs with their own "bolt holes" can prove an invaluable asset in the testing early days.

It is vital that your new dog is allowed to stake out its environment, so, if you are lucky enough to have the space, I recommend you build a separate area—a compound, if you will—in which the new dog can begin its life within the home. It should be given all the time it needs to assimilate to this space. At the same time, use the separate "neutral" area for your existing dogs to continue to get to know the stranger now in their midst—always under close supervision by you. The length of time the extended pack spends together should be increased slowly as the new member is accepted.

If your dogs get on immediately, then they can sleep in close proximity to each other from the very start. If they seem reluctant, however, they must be kept separated; good friendships blossom over time, so don't despair. The same applies with other household pets—cats, rabbits, or whatever. If they are getting along happily, then leave them together; if there is friction, separate them. Always err on the side of caution.

The Doorway to a New Understanding: Reuniting

The first of the four elements of my method that you will need to tackle is reuniting after a separation. Its importance can't be overstated; it is the cornerstone of the method, the key piece of communication between dog and owner. Yet this first hurdle is also the one that trips up many owners. In fact, in the years since I first evolved my method, no other element has been the subject of quite so many questions. This is the stage where most owners really have to overcome their human instincts and remember that they're dealing with another species. It is vital, then, that you take time to grasp this element properly. The aim of this section is to guide you and your dog through the process in simple, step-by-step detail.

I have already stressed the importance of mental preparation and of learning to think of the dog in a different way. Going into this process, you must see yourself as a member of a pack that, for the moment at least, your dog believes it is responsible for. The dog is programmed to live in a society where leadership is constantly under review. So, when the pack has been separated and is about to reunite, the dog will now need to know one thing: who is the boss now? This is your first opportunity to seize the reins of power.

I often find that owners get confused by what precisely constitutes a separation. Clearly, if your dog remains at home while a member of your family leaves the house to go to work or school for the day, or heads off to the supermarket for a couple of hours, separation has occurred. But what about when someone pops out of the living room into the kitchen, or dashes upstairs to the bathroom?

So as to clarify what I mean in my method, a separation occurs when one member of the household leaves the dog and, in so doing, erects a physical barrier between them. At this point it has become impossible for the two to communicate or connect directly. So it is not a separation when someone moves from one room to another within a house while leaving the doors open. The dog still has access to that person and can follow. If the door is closed, however, it cannot. It follows then that the first separation will occur when someone within the house leaves the dog and closes a door behind them. It is when you return to the dog's space, when you step back through that door, that you must begin to put the first principles of the method into practice.

The aim of my method is to establish a home environment into which the dog fits as naturally as every human member. My goal is to lead a dog to a situation where it is so attuned to the structure of this environment that it sees nothing unusual at all in the members of the household coming and going. The dog should accept that everyone gets on with their lives without the need to acknowledge it. It should understand that, at times, its owners are busy with other matters. There is nothing negligent or uncaring in this; in fact, quite the opposite is true. By interacting with your dog when you are ready and able to do so, you are giving your pet quality time. You will be able to shower your dog with all the love and affection it wants—the difference is that you will always be doing it on your terms.

Given all this, it therefore makes sense for the training to take place in as natural an environment as possible. Ideally, you should base the training around your normal routine. Life should go on as normally as possible, depending on the demands of your dog at particular times. Apart from anything else, this should make you and your family more relaxed about the whole process.

Stage One: The First Separation

Try to let this happen as naturally as possible—you want your dog to fit into the normal routine of the household after all.

Engineered situations aren't my preferred way of working; the less artificial these first separations are, the better for everyone.

Within the first few hours of the day, the moment of separation is going to come inevitably. Someone in the household is going to go to the bathroom or is going to step outside into the garden. At the same time, I understand that owners are often anxious to put the method into practice, particularly if they have made time to do this. In this case, I would suggest that, when a new dog arrives in your home, you delay the first separation until the dog has stopped its "recce" of its new environment and has settled down. Once that has happened, you can leave the dog and move into another room, shutting the door behind you. This should be done without any ceremony. There is no need to announce it by saying "I'm just popping out"; it should not be made into an issue. By the same token, when you return to the room, you should do so quietly. Remember, the idea you want to communicate to the dog is that, as a leader of the pack, you are free to come and go as you please.

Stage Two: The Repertoire

One of two things will happen when the dog sees you have returned. It may ignore you completely; far more likely, it will go into some form of routine, a repertoire that you should recognize for what it is: the dog's attempt to establish its place in the scheme of things. It is imperative that you do not acknowledge or interact with the dog in any way, shape, or form.

The following moments represent the first test of both you and your dog's determination to succeed. As I have explained, my method evolved as an alternative to the traditional forceful training regimes. Of course, you could at this point use physical force to let the dog know who is boss; however,

I believe we are capable of using intellect and human resourcefulness to outmaneuver a dog. We do not need to be bullies or brutes to gain the dog's cooperation. Remember: "A man convinced against his will, is of the same opinion still."

I have seen dogs go through all sorts of repertoires in their attempt to prevail in this situation. They will leap and bound around the room, jumping on furniture, they will bark and wail and drag their favorite toys into the middle of the floor. Whatever particular act the dog treats you to, it must be ignored. By making eye contact, touching, or even simply shouting "Stop it," you are conceding defeat and acknowledging your dog's primacy. You will have confirmed the dog's belief that it is the leader.

After a while, the dog's repertoire will draw to an end. The intensity of its activity will slowly fade away. Eventually, it will either wander off or lie down on the floor, as if to signal its concession of defeat. The length of time this will take is utterly unpredictable. This can be a considerable test of your determination. Dogs are amazingly resilient and determined creatures. I have seen dogs continue with this routine for anything from a few seconds to, in the case of one particularly difficult dog, three-and-a-half hours. The average time, however, is just ten minutes. The good news, too, is that the first time is by far the worst. Once the principle has been established, it will become easier and easier to re-establish leadership. It is this first occasion that places the greatest demand on an owner.

It is an area that many owners have difficulty with, for all manner of reasons. The most common mistake people make is to assume the repertoire is over prematurely. Often, dogs go through a ten-minute routine then stop before starting again for another five minutes. Owners often mistake the break they are taking for the end of the routine.

A key signal here is the dog's body language. If your dog lies down on its side, as it would when falling asleep, it has finished its routine. But if it simply drops back onto its hind legs, Sphinx-like, with its ears pricked, its language is plain: the dog hasn't finished and is ready to spring back into action at any moment.

Many owners don't succeed in this area at first, but it should not be viewed as failure. I actually believe it should be treated as a positive event, because what must now happen is that the whole process must begin again. If you fall for this "sucker punch," you should not beat yourself up about it. You should see it as a fresh opportunity to communicate the message to your dog and go back to the beginning of the reuniting routine. And this time your dog will get that message more quickly.

Stage Three: The Five-Minute Rule

You must now wait five minutes. Let me state here and now that I am not a major shareholder in a stopwatch manufacturing company—I do not want to be responsible for creating a world of clock-watching dog owners. Five minutes is the minimum recommended period of time; it can be as long as you like. As the method becomes ingrained in both you and your dog, it will be dramatically reduced.

The crucial point is that, during these initial stages in particular, your dog must be given time to absorb the powerful new information with which it is being provided. If it remains calm and relaxed for this period of time, you can capitalize immediately. An important breakthrough has been made. The dog has understood there has been a fundamental change in the power structure. It is vital that you now strike while the iron is hot and move on to the next stage, introducing the first basic control, "the come."

Developing "the Come"

At least five minutes after your dog has finished its repertoire, making sure you have not separated in the meantime, you should invite your dog to come to you. Once again, you should not set yourself up for failure here. In the early stages of my method, you should remember to have a food reward available whenever you are around your dog.

If necessary, you can squat or kneel at this stage. It is important that eye contact is made and your dog is called by its name. If it is a new dog, it may not know its name yet, but the word association will quickly be made if it is repeated. The invitation you make should be warm rather than authoritarian. It should be the voice of someone inviting over a friend rather than the bark of a barrack-square sergeant major. If your dog is reluctant at first, stretch out to it, extending your arm with the food reward visible.

It is vitally important to ensure at this early stage that the food reward is presented on your terms. The sight and smell of food may well bring a dog scurrying over automatically. Dogs are opportunist eaters, remember, so premature approaches like this must be ignored. If your dog persists in trying to get at the food, remove the tidbit from sight or, if necessary, move away from the situation and begin again a few minutes later when the dog has become calm once more.

Dogs must learn from the outset that the food is not automatically theirs to enjoy. I call it a reward for good reason: they must earn it. Another key thing to watch out for here is that it is important that your dog accepts the food calmly. If the dog comes over, it should be rewarded with the food, given warm but quiet praise, and stroked or ruffled on its head, shoulders, and neck. This neck area is hugely symbolic as it represents the dog's most vulnerable

spot, something a leader would naturally recognize in the wild.

If your dog comes over and jumps up, or rushes at you, you must get up and walk away. You must react in the same way if your dog gets overexcited and flops over onto its back, begging to be tickled. Many owners find this difficult, but your dog must learn the consequences of its actions at this early stage. It must accept your status as the leader—and one of the leader's privileges is deciding how, when, and where affection is given out. If your dog doesn't respond to the come at all, you must also walk away. In all these cases, your dog should be left alone for an hour or so before the routine is begun all over again.

This approach may seem harsh, but it is far better to turn around and restart the journey now than take this wrong turn and continue down a road that will lead to major problems later on. If you long to fuss over your dog, all you have to do is get up from where you are, walk across the room, and call your dog to you. If it complies with this, you can make all the fuss of the dog you like!

Your goal is that, by the end of the first separation, your dog should have learned to relax completely, come to you on request, and accept a food reward without leaping in your face. Of course, it is not always going to happen at the first attempt.

How to Know When Your Dog Has Accepted You as Leader

Dogs will, of course, manifest their personalities in different ways. Some will signal their relaxed state in obvious ways, by lying down or playing with a toy, chewing on a bone or quietly grooming themselves. But you can also look out for subtler signals.

The Eyes Have It: A key technique for spotting that your dog has relaxed lies in its eyes. If your dog is staring wide-eyed, then it is still in an anxious state; it has not accepted what is going on. If, on the other hand, the eyes are soft, then it is relaxed and ready to be called over.

A Bridge of Sighs: An important indicator of a dog's level of anxiety is its breathing rate. If it is breathing heavily and rapidly, it is still stressed, so wait until the breathing rate has steadied. The telltale sign that it has relaxed is when it releases a long, deep, peaceful sigh. In humans, a sigh is an ambiguous signal; it can mean frustration, boredom, or defiance. In this situation with dogs, however, I always interpret it as a release of stress. A weight has been lifted off your dog's shoulders, and it is relieved to see it gone.

This relief can, in some cases, manifest itself in the most graphic way. I have seen marked changes in the toiletry habits of some dogs. They urinate and defecate in much larger quantities. In the case of the urine, it is easy to see why this is happening: while the dog retained its status as leader, it instinctively retained large quantities of its urine for territorial marking purposes. Freed from the responsibility of leadership, it empties its bladder in full. I must confess I was taken aback when I first encountered this. It also made me realize how "bottled up" dogs are when they reluctantly cling to the reins of power within their pack.

Stretching Out: Some dogs will go through a routine of stretching, much like a human might do getting out of bed in the morning. They will arch their backs and splay their legs out in front of them. This is a positive sign. It means the dog is relaxing.

Lip Reading: As they weigh up the situation, some dogs will begin licking their lips a lot. This is a positive rather than a negative sign. We humans do this when we are apprehensive or deep in thought; in the dog's case it is a signal that it, too, is thinking, probably wondering, "What's going on here?" The important thing is to let your dog arrive at its decision. It will get there in its own good time.

There are, then, many indicators. But, if you are still unsure whether your dog has relaxed, you can carry out a simple test by moving from where you are sitting to another position. If your dog is relaxed, it will not react to this. If it makes a move, more time—and more patience—is required.

Frequently Asked Questions

What's Wrong with My Dog Being Leader?

The simplest way to answer this much-asked question is with another question: would you hand over the responsibility for running your household to a one-year-old child? I often say that we keep dogs as puppies for life. By this I mean they never learn to fend for themselves in any meaningful way. They are, in effect, permanent one-year-olds, yet many, many owners inadvertently tell these immature animals that they expect them to perform the role of leader of the entire household.

This situation is clearly disastrous for all sorts of reasons. The dog has no way of understanding our society or language and, by elevating it to a job it is utterly unequipped to deal with, the owner places it in an awful position. It goes out into a world filled with a thousand threats, unable to cope with any of them. The result of this is that the dog feels a failure. And from there a vicious circle begins: as it fails to lead effectively, it gets more stressed and its behavioral problems are going to begin.

People often say to me that their dog has a natural air of leadership and may be frustrated or unfulfilled by being "demoted." Of course, some dogs do have a more dominant air than others; if these dogs are members of an extended pack, they will tend to head the subordinate pecking order within that pack. But even these types of dogs are happiest in an environment they understand, running free in open spaces, eating, sleeping, and playing within a safe and secure pack environment.

What if My Dog Misbehaves Because It Is Being Ignored?

Some hyperactive and aggressive dogs can work themselves up into a terrible state when they are given the cold shoulder in the way I advocate. They are used to demanding and getting attention. When this is denied them, they react according to their natures by selecting one of the "three Fs"—flight, freeze, or, in the worst-case scenario, fight. They can bark loudly, jump up into your face, or even bite. Introducing my communication method to such dogs requires a measure of planning and perseverance, but the results will be just the same.

If you have a dog with a strong personality like this, you should think about putting a collar on them before beginning. You should also prepare a room or an area of the garden where the dog can be safely excluded if necessary. In this respect, planning the geography of the training area is a classic example of being properly prepared before beginning. I advocate using a room for training and another room as an exclusion zone. If you think there is the remotest chance of your dog becoming hyperactive and "trashing" the room, then you must clear out the room before starting, so as not to set yourself up for failure.

When the first reunion occurs after separation, if your dog reacts in an unacceptable way, you should take it by the collar and lead it away from the room to its exclusion zone. This is a crucial moment in the struggle for leadership of the pack, so it is vital that you demonstrate all the qualities of leadership at this point. You must not shout, you must not get upset, and you must not speak to or acknowledge your dog in any way. Your pulse rate must remain low, and you must act firmly and decisively.

When you put your dog into a room or out into the garden, you must shut the door behind it. Eventually the dog will calm down and be quiet. Ten seconds after it has fallen silent, you should open the door and let your dog back into the main room. If the dog repeats the behavior of earlier, you must immediately lead it back into the exclusion zone, shut the door, and wait for it to become quiet again. This time you should wait a longer period, say thirty seconds. If your dog is still quiet after this time, let it back into the room. If the bad behavior starts again, repeat the process. Each time this is repeated, your dog should take a step in the right direction. The intensity of its barking or bounding should diminish, and the length of time it remains quiet while waiting to be released from the exclusion zone should increase.

In really extreme cases, for instance if your dog bites, I suggest you attach a lead to its collar. The dog should be led out from the main room to the exclusion zone at a safe length. If a dog is so agitated it is difficult to even lead it out of the room, I ask owners to reverse the process and to leave the room themselves. You must then wait until ten seconds or so after the dog has quieted down before going into the room. The instant the bad behavior is resumed, you must remove yourself again, returning this time thirty seconds after the dog quiets down. As before, this process must be repeated until the dog has conceded defeat and relaxed.

It can be very hard work. I have worked with owners who have had to go through this routine seven or eight times, but I cannot overemphasize how important it is to remain determined. The lessons your dog is learning at this point are life-changing. By the end of the process it should have redefined its status within the domestic pack and, at the same time, it should have understood the consequences of misbehaving. We are attempting to stage a peaceful revolution here. By overcoming this particular hurdle, you will have made the crucial first move and will have stormed the barricades—by stealth.

What About Nervous Rescue Dogs?

Too many people forget that dogs are living, breathing, and—most of all—feeling creatures. Like us, they develop according to their own abilities, and we must be prepared to accommodate that into our lives. We are not programming a video recorder here, we are asking creatures to make adjustments to their ways of thinking. So it follows that getting them to understand the powerful signals within my method is something that is going to take more time in some cases than in others. And in the case of some rescue dogs, this can take a significant amount of time.

To my mind there are no such things as "problem dogs," but there are dogs with problems. Rescue dogs have come to sanctuaries or rescue centers because they have been mistreated by humans in some respect. In some cases, of course, they have simply been abandoned. In the more extreme cases—and there are more of them than I care to think about—their experience of life with humans has been wretched. A dog may have been mistreated for years before it is rescued, and the damage done during this period is not going to be undone overnight. For this reason, you should think of these dogs as animals with learning disabilities—sometimes severe ones. You must realize that, while the dog will learn, it is going to do so at a slower rate. And it is going to need that extra bit of love, patience, and understanding to get there.

How Can I Include Young Children in the Method?

Young children are clearly not going to be able to grasp the principles of my method instantly. In time, as the rest of the family masters the process, they will see how things work and be more receptive to copying what their parents, brothers, and sisters are doing. It is going to be a gradual process.

Having stated this, however, the smallest members of the household can make a contribution to correcting a dog's mindset. Children are never alone in the house so, invariably, the major reunions of the day will involve the children arriving home with you. They may be returning home from the shops or school, and this is where you can dispense a double dose of medicine. By steering your children away from the dog when they first enter the house, you can engineer the situation so that the children are also withholding the acknowledgement and homage that the dog expects. If, for instance, a child is placed in its high chair and distracted by a toy, a drink, or some food, it will be oblivious to the dog's attention-seeking repertoire.

If, once the five-minute rule has been applied, the child is then involved in calling and rewarding the dog, an important statement will have been made about the status of this member of the family within the domestic pack. Obviously, this is not a situation that is going to be possible at all times, but it will do no harm to establish the idea in the dog's mind.

When Should I Next Separate from My Dog?

Separation is an intrinsic part of the life you and your dog are going to lead together. On the first day, however, I ask owners to minimize the amount of times that they separate from their dogs. After the first successful come, for instance, you can continue calling your dog to you while you remain together. You can also build up so that you begin playing with the dog too. Your desire to bond with the dog—especially if it is a new arrival in the home—will be considerable and understandable. The only thing I ask is that you are cautious with the amount of affection you give your dog during this very early stage. It is all too easy to drown a dog with too much affection, making it hyperactive, needy, and potentially problematic further down the line.

It is also not necessary for you to interact with your dog after each and every separation. By simply ignoring the dog and going on with your daily routine, you are still relaying the central message that you are in charge of its movements. The fact that the dog is not invited to join you will further underline the message that it is you who dispenses affection and social activity within the pack.

It is also likely that other members of the household will be entering the house, and they will hopefully underline this message. Again, this is where preparation is so vital. With the dog now established in its den, all newcomers to that den must now assert themselves as a senior member of the pack immediately. They must start as they mean to go on. This, I realize, is going to be particularly hard when a member of the family comes home to meet a new dog for the first time. Hard as it may be for them to ignore the bouncing ball of fun that has just arrived in their home, they must do so until it quiets down. Like everyone else in the house, once they have applied the five-minute rule, they are free to welcome the dog to the house with affection and praise.

When Can I Stop Ignoring My Dog Each Time We Reunite?

The answer to this commonly asked question is: never. My method is not a training scheme that is dropped when it has achieved its desired effect; it is a lifestyle. If you begin acknowledging your dog when you reunite, the dog will soon claim leadership. This must not happen.

At first, many owners are daunted and discouraged by this prospect. What they soon realize, however, is that three things will happen within a short period of time. Firstly, the routine they have begun to practice will become second nature. They will automatically ignore the dog on first being reunited with them. Secondly, and more importantly, the dog's repertoire of attention-seeking behavior will fade away. As a result of this—finally—they are able to reduce the amount of time they have to wait before calling the dog to them. Gradually, the five minutes is reduced to mere seconds.

I understand why it is difficult for owners to see this in the difficult early days. But I promise that, fairly quickly, you will be able to return to the house after a day's work and, by the time you have walked past the dog, carried on to the end of the hall, and taken off your coat, your pet will have completed its ritual and will be relaxed. So well established will the routine be by now that you will then be able to immediately call the dog to come to you and reward them with all the praise and affection you like.

Walk Tall: Learning the Art of Leadership

"THE LEADER MUST KNOW, MUST KNOW THAT HE KNOWS, AND MUST BE ABLE TO MAKE IT ABUNDANTLY CLEAR TO THOSE ABOUT HIM THAT HE KNOWS."

Clarence B. Randall

My method requires all owners to assume the leadership qualities their domestic pack demands of them. Time and again, however, I have seen that this is an area where many come unstuck. Many people find it hard to impose themselves as the dominant force in the household. There can be all sorts of reasons for this: some people are simply shy and lacking in confidence; others are gentle souls who just aren't used to taking the lead or expressing their personalities in such a forceful way. Yet, in every case, I believe a little knowledge can prove a powerful thing.

By demonstrating the five key qualities below, all owners can quietly and efficiently take on the mantle of leader within their own particular pack.

Walk Tall

A great deal of leadership is down to projection. Body language is a means of communication that dogs understand all too well. They will pick up on anyone who seems tentative or weak immediately. Equally, they will respond immediately to someone who displays obvious signs of authority. As we all know, first impressions are lasting impressions, so it is vital that you present the right image from the very first minute of the first day. Walk with an upright, confident posture. Keep your head high, and your eye level above that of the dog. Avoid any eye contact until you are ready to interact with the dog.

Be Decisive

A leader's actions should be firm, final—and immediate. Your dog will test this from the start, so it is vital that you are not thrown by behavior like gesturing or barking. You must continue to ignore the dog until you are ready to acknowledge its presence, and rebuff it immediately if it comes to you uninvited. Even allowing it to rest against your leg for a few seconds allows the dog to believe it has some authority.

Silence Is Golden

We have all met owners who talk incessantly at their dogs: "Good boy, here boy, that's my boy." This is counterproductive, as the constant noise becomes nothing more than a background drone. Dogs respond much better to a leader who speaks only when they have got something important to say.

Speak with Authority

It is vital that your tone of voice is consistent with the body language you display. I ask owners to speak clearly, confidently, and decisively. Short phrases or words are best: "Come," "Sit," "Stay," all convey powerful yet simple messages. Think of the most effective leaders in our lives: none of them minces their words. Deliver praise with a smile and an extra softness to the voice, using brief phrases like: "Good dog," "Good, clean dog," or "Clever dog." My method is not so cold and mechanical that there is no room for affection either. I sometimes can't resist saying: "I do love you."

Believe in Yourself

Of paramount importance in my method are the "two Cs," being calm and consistent. To these two requirements, however, we need to add a third C. You must be convincing to the dog, and this is not going to happen unless you believe in yourself.

Many dog owners are beaten before they begin. Rather than convincing themselves that they know what they are doing, they convince themselves of the opposite. In their minds they think success is beyond them. Often, my most important job is to persuade the owner that the goal is not as impossible as it appears. Invariably these people are convinced they are simply not "leadership material." They are certain they are not strong enough to take charge of a situation and to have their pet believe in them. A typical example I have worked with was Anne, a single girl, in her twenties, living alone with her schnauzer, Teddy, in Kent.

Anne called me because she was having problems with him. Schnauzers have big personalities, they have a lot of attitude. Teddy was running away a lot and was pushing Anne around at home. When I met Anne, I saw that she was a small person, just over five feet tall. She was also slight in personality and was a very nervous, insecure person. I sensed that Teddy was aware of this too. It was clear almost immediately that, as long as she remained like this, Teddy was never going to relinquish the role of leader within their domestic pack. As I chatted to Anne, I explained that the onus was on her to change things. "For the first time in your life I want you to take charge," I told her.

There are, of course, physical reasons why some people find it difficult to impose leadership. Bigger dogs are powerful animals and owners who are small or frail can find them hard to deal with. Yet there is no reason why owners cannot behave in an authoritative way. If I am helping someone who finds themselves at a physical disadvantage with their dog I encourage them to walk away. If the dog invades their space or tries to impose itself on a situation, I tell them to leave the room and shut the door behind them. The response must be decisive and instant. If the dog sees its attempts to manipulate the situation are a total failure, it will quickly get the message.

In Anne's case, this was not necessary. She was physically capable of dealing with Teddy; it was her leadership qualities that she needed to work on. My task is often to change the way both the owner and their dog think. It is not my place to make them follow my method; I must make them want to do it. If the owner wants to change the way they communicate with their dog, then the knock-on effect will be that the dog will want to live that way too. It is one of the key areas to unite us as species; as long as there is something in it for us, we will go along with a situation.

We set about working in my normal way. After applying the five-minute rule, I asked Anne to call Teddy to her. I asked her to do it in an inviting way. At first she was tentative. So I asked her to say it again as if she really meant it. She asked Teddy to come, but this time she said it boldly—there was warmth and strength in her voice. Teddy looked at her. He sensed there was something different and came to her almost instantly. It was a promising beginning.

When I go on a call, I am putting the brakes on a downward spiral. Success breeds success and, when you get success, you become more confident. The dog realizes you have that confidence, which leads to more success and more confidence and so the upward spiral begins. A month down the line, Teddy and Anne were doing fantastically well. I could hear from Anne's voice that she was gaining confidence. When I spoke to her three months later, it was the changes in Anne that again struck me the most. She was speaking with so much more confidence. She was still a quiet, self-effacing woman, but at the same time she seemed more sure of herself. We ended up having a long chat, and it emerged that she had started to socialize and had made a friend of another woman who walked her dog. "It is all to do with the belief I now have that I know what I am doing," Anne admitted.

I must admit I became very attached to Anne and we

spoke again over the following months. She worked as a sales person for a pharmaceutical company. At one point she told me she had moved on to a better job within the industry. Again, she had become aware of her abilities at her work and had reaped the benefits of this self-knowledge. What I now realize is that Anne's need to communicate was as strong as her dog's. She has not become an extrovert, the personality of the year, but she's certainly become a lot more confident and more outgoing. It was wonderful to be a witness to this transformation.

Family Reunions: Reuniting with Multiple Packs

If there has been one dominant theme in the questions I have fielded since the publication of my first book, it is the issue of introducing my method to more than one dog. So many people now seem to have a number of dogs, and so many of them have asked me about this that I feel I should devote some time to their concerns. I am not surprised at their worries. Quite naturally these owners assume that, because the dynamics of a pack of dogs are so much more complicated, the method must therefore be equally convoluted. The good news is that the principles underpinning my method remain exactly the same. The even better news is that teaching a number of dogs together is, to my mind, easier than working with a solo dog.

Essentially, you have three options available to you. Dogs can be taught either individually or as a group, or larger packs can be taught with a blend of both—as a series of groups or "mini packs." To decide which of these suits you best, you must be aware of the relationships your dogs have with each other. The politics of the pack is going to be an enormously useful tool and, for you to understand them, you should have an understanding of the natural society from which your dogs have come.

Pack Awareness

Each pack is unique. No matter what the mixture, whether they be dogs of the same breed or a blend of different dogs, the web of relationships they contain is different. For this reason, you must be aware of and understand the dynamics within your pack, to avoid being caught out. By underestimating the forces at work within a pack, you run the real risk of setting yourself up for failure.

Ultimately, the forces that drive a pack are no different from those that drive an individual dog; survival is the ultimate priority for both. Once more, the roots of those instincts lie within the dogs' ancient ancestors and community, the wolf pack. For millennia, the wolf has steadfastly accepted the hierarchical principles that underpin pack life. The Alpha pair command complete respect. Their primacy is beyond challenge and the rest of the pack feel safe in the knowledge that the pack is their best option for survival.

The remainder of the pack has its own ranking system, ranging from second-in-command Beta males and females, to yearlings and pups. Each has a clear grasp of its job description, whether it be joining the Alphas on the hunt or remaining in the den on guard duty. The complex variations need not exercise us too much; the key point to understand here is that the lower-ranking wolves are all perfectly happy in the subordinate roles they occupy beneath their leaders. As long as the pack survives the tests of everyday life, the wolves are content with the ranking they have at this lower level. It is akin to a successful corporation run by a charismatic, intelligent, and hugely resourceful chief executive—Richard Branson is the obvious example here in the UK. The corporation's staff believes in this leader totally. Their lives depend on the company's success and they know he's their best chance of delivering it to them. They believe in the power structure that has been assembled below him because it works.

But a company does not become successful by resting on its laurels. And it will not survive, let alone thrive, if it takes anything for granted. Therefore, the leader is constantly being observed by his staff, his mettle is perpetually being tested. The leader knows this, and uses the fact that he is constantly being judged to his advantage. He knows that, as long as he continues to exude an air of confidence, that confidence will seep through the ranks. There will, of course, be rivalries and allegiances within the ranks, just as there are in any company—not everyone is going to get along. And if the leader is lost, those rivals will battle it out to succeed him. But, unless the company's existence is threatened, everyone is more than happy to abide by the pecking order. The status quo will be maintained. It suits everyone.

This is how I ask owners to think of their domestic pack. Dogs are naturally submissive creatures—they will do anything for a quiet life. It is why they allow humans to bully them so easily, but is also why they are so willing to be led by others. A pack is at its happiest when there is a sense of natural order, a harmony among the dogs. They will test the leader particularly at the four occasions when the pack's survival is on the line: when reuniting after separation, at feeding time, at times of danger, and during the hunt. Provided they are confident in the abilities of their leader to deal with these situations, they will fall into line behind them.

It is important to realize, however, that each pack will have its own dynamic. Dogs are, as I have emphasized before, living, feeling creatures. So, inevitably, there will be alliances and rivalries, some dogs who are close friends and others who simply don't get along. This is easy to detect; dogs that dislike each other will turn their backs on each other or avoid eye contact. They will keep a distance between themselves, and take positions at different ends of the room—their body language is obvious. There will also be occasions when these tensions are more likely to surface. These are usually connected with times of uncertainty such as when new dogs arrive or old dogs leave, or when members of the pack are in season.

Some other alliances will be equally easy to read. Blood ties will obviously be tight, but that does not mean there cannot be sibling rivalries too. Equally, there will be some dogs that are more dominant than others. It is something I have seen in all my packs. In my first book, I talked about Donna, the dog we called "The Duchess." She expected everyone, human and canine, to defer to her; she was a rare, natural Alpha. This is often a matter of sheer personality. The crucial thing to realize here is that all these things are within a dog's nature and we cannot do anything about them. The one exception to this is, just because we cannot fight nature, does not mean that we should let dogs fight; violence and aggression between pack members must never be allowed. In general, the pack's overriding instinct is to preserve the status quo, to keep things as they are. They know they are better off together, that there is strength in numbers and in the concept of community. These are powerful forces that can be used to your advantage.

Teaching a Pack as Individuals

Every owner's circumstances are, of course, going to be different. It simply may not be possible for you to cope with the process of working two or more dogs at the same time. Equally, there may be particular tensions within the pack that prevent pack work. Clearly, for some owners, the best option is going to be working with the dogs individually. The key thing here is that, by the end of the process, the pack has been brought together. This way you will benefit from all the positive interactive forces within the pack.

You should begin the process by isolating one dog at a time from the pack and going through the initial technique of reuniting after a separation. If one dog within the pack is particularly problematic, disruptive, or simply more high-spirited than the rest, I would recommend leaving this one until last. Certainly in the case of rescue dogs, which are generally highly nervous animals, this is by far the most desirable way of beginning. My reasoning here is simple and based on personal experience. If a difficult dog rejoins the pack to discover its peers in a relaxed and cooperative mood, it becomes far more receptive to the idea of joining this responsive state. The influence it sees you wielding over its pack colleagues is hugely powerful. By the same token, the dominant dog should be worked first if possible. The knowledge that this dog has accepted you as leader will be a hugely powerful weapon with the rest of the pack.

When you are ready to begin reuniting work, your first dog should be disregarded in the way I have already described. Again, it is important to stress that this process cannot be rushed. Dogs will surrender themselves according to their own timetable, so if you are going to attempt to work with a number of dogs in this way, you must put aside a sizable amount of time to put this into practice.

The first dog should go through its repertoire and relax entirely. Once this has been achieved, the next member of the pack should be brought into the room. Almost certainly at this point, dog number one will revert to its repertoire again. This is perfectly natural—after all, the two dogs are reuniting after a separation as well. The likelihood is, however, that the repertoire will last for a shorter period of time. The key thing to observe here is that both dogs reach the point of relaxation before you move on. Once the dogs have reached this point, the next dog must be brought in. This time all three must calm down before moving on.

This process should continue until the last dog has rejoined the pack. As I have said, I recommend that the most disruptive dog within the pack should be left until last. Hopefully this dog will enter a room in which it finds that its peers are displaying homage to a powerful new leader. It will almost inevitably attempt to subvert this process. But, by remaining calm and displaying convincing leadership, you should eventually see the whole pack reaching a relaxed state. It is now that the five-minute rule should be applied. Again, it is worth stating that this is only a minimum period; reaching this point may well be a long, hard struggle. Provided you do not separate from your pack, you may well want to take a longer break.

The Come

When you are ready to carry on, you must now build on the level of leadership you have already achieved. And the way to do this is by establishing that you have the power to choose which dog you interact with from now on. At this point, call one of the dogs to you; it doesn't matter which one it is. It is important that you only look at and acknowledge the dog you have called and ignore the others. Any dog that runs up or attempts to jump up must be steered away firmly but fairly. In contrast, the dog that has been called to come should be rewarded generously when it does so.

This process should then be repeated with other members of the pack. As each of them learns to come to you as requested, it should be rewarded in the same way. Each time one of the dogs behaves inappropriately by coming when it has not been called, it should be ignored and, if necessary, gently rebuffed by you, the leader. This is a potent time. The pack as a whole is seeing a radical new form of communication. A seismic shift in the hierarchy of the extended pack has occurred. Things will never be the same again.

If you own a particularly nervous dog, I would recommend that you leave it alone. Such dogs tend to remain disconnected, almost aloof from the rest of the pack during this process. There is nothing wrong with allowing it to stay on the fringe of things. As I have explained, it is not necessary for you to interact with a dog each time you reunite. Besides, simply observing the rest of the pack in action will act as a powerful stimulus for it.

The nervous dog will signal its readiness to interact soon enough. It is important to remember, of course, that you should not let the dog make the first move, no matter how much you are looking forward to that first affectionate moment. There can be no predicting at what time this will happen and you certainly shouldn't expect it to happen in an instant. Like all good things, however, it is worth waiting for.

Teaching Dogs as a Pack

I often say it is easier to train two dogs than one. By this I mean that there are strong forces at work within a pack of two or more dogs that can be channeled by an intelligent owner. Essentially, pack teaching involves applying en masse the individual method outlined above. The process begins with all members of the pack present. When all the dogs have exhausted their repertoire and have relaxed, individuals should then be asked to come in turn. For fairly obvious reasons, this is only going to be practical with small packs of two to four dogs. The potent peer pressure applied by their fellow dogs, along with the animal's deep-seated need to belong to a pack, are invaluable tools.

* * *

I have come across no better an illustration of this power of the pack than the case of Dish and Spoon, a pair of Jack Russell terriers who had grown from being inseparable friends to lethal rivals. Their owners, George and Estelle, had taken them into their home, an old rectory near Louth in Lincolnshire, as puppies from the same litter. As puppies, they had played incessantly—exactly as they should—using the experience of tugging and pulling and wrestling with each other as a means of developing a whole range of skills. Three years on, however, the pair's habit of wrestling with each other had taken on a much more sinister and upsetting form. They would regularly egg each other on. In the fights that ensued, they had suffered bites and punctures to ears, lips, and legs.

George and Estelle were devoted to their dogs and had done all they could to help Dish and Spoon. A vet had recommended that the more aggressive of the pair, Spoon, be castrated. I may not win very many friends in the veterinary world for saying this, but I am against castration as a means of calming aggressive behavior in dogs. The procedure obviously has its place in the dog world—in medically necessary cases and where dogs need to be stopped from reproducing. My objections tie in with my general overview of dog behavior. Castration does nothing to relieve the dog of its delusions of power. All it does is lower its testosterone level and create an even more wounded animal. The less it feels able to do the job of leader, the more panic-stricken and the more aggressive it becomes.

As it turned out, events with Dish and Spoon bore out my case. The castration did little to calm Spoon down. In fact he became more aggressive. The vet then suggested that Dish be castrated. The results were, to my mind, inevitable. One day, George and Estelle were in the garden with the two dogs when Spoon suddenly flew at Dish and launched the most ferocious attack either of them had ever seen. The spectacle of your own two dogs trying to rip each other to pieces is a shattering thing for a dog lover to behold. The dogs were left in a dreadful state afterward. Dish had a gaping hole in his chest. Spoon had puncture wounds on his face and legs. It was at this point that George and Estelle called me in.

As I listened to their story, I sensed how heartbreaking it had been for George and Estelle to see such great sibling ties broken. I explained to them that, in applying my method, they could use those ties to rekindle and revive the brothers' old relationship.

It was clear to me that the root of Dish and Spoon's problem lay in their competing wish to lead the pack. This was obvious from the times when George and Estelle told me trouble flared up most easily. Predictably, these were when people came to the door or walked past the gate outside, when there was an aircraft passing overhead, or, sometimes, when Estelle was preparing food. I knew that the powerful bond that once connected these dogs had to be used to bring them together once more.

As I began, the key thing was to let each dog see the other undergoing the process. The two dogs were separated by a gate between the kitchen and the sitting room so that, while I was working with Dish, Spoon could see clearly what was happening and vice versa. I went through the usual five-minute rule routine. I was encouraged early on by the fact that, rather than focusing on the other dog, the dog I was working with gave its full attention. Also, I noticed a distinct lessening of the growling coming from the non-participating dog as it watched the other dog respond to the signals I was giving it. Eventually, encouraged by the other's behavior, it too calmed down. I then worked on the come, with both dogs still isolated but within view of each other.

After a while, I suggested we took the two dogs for a walk on leads in the garden. George and I took them out, being careful to keep them in opposite parts of the garden. They eyed each other up for a while. Slowly, however, both relaxed, allowing us to bring them closer and closer together. It was not long before they were wagging their tails and looking at each other. "It was as if they were meeting each other for the first time," George said.

I knew there was still a long way to go, and suggested we now walk them alongside each other. Once again, I was going to use the natural power of the pack here. It was not long before Spoon gave a low grumble at Dish. I immediately led him away from the garden into a room in the house. I now wanted to use the dogs' natural need for companionship and the safety of numbers as a lever. As Spoon was removed, his body language told its own story. It was slumped, much like a sulking teenager who has just been sent to his room. Spoon was clearly hugely disappointed at being separated like this. Deep down, he had a desperate desire to be with his sibling. I released him from his room after fifteen minutes. He behaved slightly better but growled again after a while. Once more, he was banished to the room. Once more, we saw the sulky stoop.

George and Estelle grasped what was going on. They could see that, with each separation, Spoon's attitude to Dish was changing. He was realizing significant changes were afoot. By the end of my visit there, George and Estelle felt very positive and assured me they would build on this. Over the following weeks, they worked on these early breakthroughs, and soon Dish and Spoon were able to go out for short walks together in the garden. At this point, however, they were both still on the lead.

I suggested to George and Estelle that the first occasion they let the two dogs off the leash together should take place on neutral ground. Six weeks later, I got a phone call from them asking if I'd be willing to join them when they released the two dogs in a friend's garden. Of course, I was pleased to be there.

On the big day, George took one dog while I took the other. Slowly we lowered the leads to the floor and let them go. To our delight, the two dogs just carried on investigating the garden, heading in different directions. I remember George gave my hand a squeeze, but I quickly cautioned him against getting too optimistic.

In the weeks that followed, Dish and Spoon could be left off the lead without any confrontations. Things have improved steadily from there. The old friendship has been restored, yet common sense has had to prevail too. As a sensible precaution, George and Estelle now always separate them whenever they get too exuberant. And they always make sure the pair are separated when they are left on their own together at home. These small sacrifices have been more than worthwhile, however. To see the siblings together and happy again after all they have been through is a joy.

Group Therapy

The third option for introducing my method applies in larger packs, where, for purely practical reasons, owners may find it easier to work with smaller, more manageable groups of dogs. This approach may also suit some owners who have a mixture of puppies and adult dogs. Given the high energy levels of the puppies, it may be easier to work with the younger dogs separately from the older dogs.

Whatever the reason for the decision, the method should be the same. Take each group and teach them together, in the way outlined above. When you have achieved success with both groups separately, you can then move toward gradually integrating the whole pack. This is not something that should happen in one move. If you feel happy and in control, move on—if not, don't. You should integrate the various factions and family groupings slowly, again using the pack dynamics to work in your favor.

Pack Problems

New Dogs and Existing Packs

It is, as I have already explained, imperative that you are careful about introducing new dogs—and in particular, older dogs—to existing packs. Regardless of how well attuned the existing dogs are, the arrival of a new personality can be the spark for disruption, tension, and outright aggression. In the worst cases, this can have tragic consequences.

I recommend you only introduce a newcomer once you have successfully established yourself as the unchallenged leader within your existing pack. The pack-training option you use will be entirely dependent on the dynamics of the pack, of course. I would also advise that you introduce your most easy-going and friendly dogs to the newcomer first. If you have the space, work in the separate, "neutral" area where your dogs have been getting to know each other.

Lengthy Separations

Owners of packs need to be particularly careful when the pack is split up and separated from now on. In their natural environment, dogs are simply unused to members leaving the pack for any length of time. Short separations are far less problematic, but if a dog leaves for an extended stay at the vet, or travels with you whilst the other dogs are left at home, the housebound dogs will assume the pack dynamic has changed permanently.

A good comparison might be in the workplace. Imagine someone leaving their job to join another company: their departure would cause a restructuring. What if they were then to simply show up again, expecting their old position back? This, in a nutshell, is what the pack is faced with after a lengthy separation. (In the canine case, the matter is complicated further by the fact that the returning dog arrives back imbued with strange new smells that are unfamiliar to the pack.) So the situation must be approached with great care.

What invariably happens in these situations is that the dogs that have remained at home will go into a repertoire with the dog that has been outside. To minimize the amount of aggression that occurs, I recommend that the dogs are reunited one at a time—always ensuring that none of the househound dogs are left on their own. If a dog is left on its own in this situation it will feel it is being isolated and therefore punished for no reason. So, if there are three dogs within the pack, bring both in to meet the other when the walk is over. If there are four in the pack, bring one and then two in. If there are five, bring in one, then another, then the final two—and so on.

Training a Litter of Puppies: The Goxhill Mafia

On the face of it, the prospect of training a litter of lively puppies seems a daunting task. Controlling even one bundle of hyperactive energy appears challenging enough; attempting to impose your will on six or seven of them seems like mission impossible. I have had to deal with all manner of packs in my time, including puppies. None left an impression to compare with that of the half dozen, five-month-old, cross Border collie puppies I was called in to treat in the very early days of my work. Their case demonstrates how my method can prove remarkably easy, not to mention fun, to apply.

Puppies can bring chaos even to the most well-organized homes. We're all familiar with the image of the unraveled clothing and the unfurled rolls of toilet paper tumbling down the stairs. But the very special brand of pandemonium this bunch brought to their owners, a family of six, in the village of Goxhill, Lincolnshire, almost beggared belief.

On the telephone, I had wondered whether the owners were exaggerating. They described how, at regular intervals, the puppies would sweep through the house like a tornado, leaving a trail of destruction in their wake. No sooner had I stepped through the door than I realized that the family had, in fact, been downplaying the dogs' devilish tendencies. I called them the "Goxhill Mafia."

The owners showed me how the kitchen bore the brunt of the storm. The puppies would help themselves to food, chomping their way through packets of biscuits and cereals; they had even bitten their way through tins. Any plates, pots, and pans that stood in their way were brushed aside. Nothing, it seemed, escaped their juvenile jaws.

Over the last few months, the calamitous puppies had bitten their way through floorboards, copper piping, and even an electrical cable (fortunately it had been disconnected). The family's clothing had been reduced to a pile of rags. The garden outside was a scene from a post-nuclear holocaust. The family had an elderly Yorkshire terrier; such was the mayhem the puppies were causing, he had spent three months permanently holed up upstairs. Try as the owners might, he would not venture down into the newcomers' territory.

To their eternal credit, the family was determined to find a way to integrate the puppies into their home life. They had decided to have the puppies, and felt it was their responsibility to ensure they were safe and cared for properly. Fortunately for myself and the family, the puppies' mother remained at home with her litter. Anyone tackling a pack of puppies must begin with the mother first; to her brood she exerts a powerful natural authority. If she is willing to relinquish responsibility to a new human Alpha, her innocent youngsters will swiftly follow her lead.

I worked with the mother on her own to begin with; her protective instincts would have been in overdrive if I had been working with her children at the same time. She presented no real problems and was soon relaxed. It was at this stage that I asked for the puppies to be brought in one at a time.

Separating the pack proved a real test in itself. It took three members of the family to keep the remaining five puppies on the other side of the gate in the garden, while the first of the litter came in. Predictably, the puppy went crazy when it was brought into the room with its mother. The mother displayed a little anxiety at first, but soon remembered the newly established order. It took about five minutes for the puppy to follow its mother's example. One down, five to go.

I then asked for the next puppy to be brought in. Again, the dog went into overdrive when it arrived in the room. And, as invariably happens, the first puppy resumed its leaping around too. But this situation did not last for long and, within five minutes, the second puppy had picked up on the first's example and was also nicely relaxed. I carried on this way with all seven dogs. It took roughly the same length of time to bring each of them under control. Each time a new puppy entered, the other dogs went back into their old, frantic repertoire before the lights came on and they recalled the new routine.

It was a classic example of dogs coming to a decision they knew would be of benefit to them. They identified me as a competent, convincing leader and, led by their mother's example, were happy to invest their trust in me. It was, as it should always be, no decision at all. Why be burdened with the responsibilities of leadership when someone perfectly qualified for the role was there ready and willing to take charge? In effect, I had made the Goxhill Mafia an offer they simply couldn't refuse.

It was a magic moment. Surrounded by seven placid dogs, I looked up to see a group of people with their jaws on the floor. The family's sense of disbelief deepened when a friend of one of their children arrived unexpectedly. Ordinarily, this would have been the signal for the puppies to explode into action. Instead, they looked up at me briefly, then returned to their new, low-key routine when I thanked them for their help. I must admit I felt rather pleased with myself at that point.

The test then was for the family to apply the method consistently at home on their own. It took them a month of hard work—working with all the dogs in the home, not just the puppies—but they got there. After another month, even the Yorkshire terrier came downstairs. And the Goxhill Mafia, as far as I know, scared no one from then on.

Food Power and Gesture Eating

I am a great believer in simplicity and the idea that less can often be more. So often in life, I have seen people making the mistake of overcomplicating things. It is why I work hard to keep my method as straightforward and uncomplicated as possible at all times. The simplest part of my method—and by far the easiest for people to grasp and put into practice—is what I call "gesture eating." It is a classic example of less sometimes being more. For all its simplicity, it may well be the most powerful signal of all.

As I have already explained, the distribution of food plays a crucial role in the life of the dog's ancient ancestors, the wolf pack. It is hardly surprising that the wolf Alpha pair are at their most dominant during the gathering and consumption of food. Their job, after all, is to preserve and protect the pack; without them, the pack would not survive. Meal times enshrine this order in the most symbolic way. By ensuring they eat first at meal times, the Alpha pair signal in unequivocal terms that they are the leaders. By giving their leaders first refusal at every meal time, the rest of the pack are acknowledging that they instinctively understand this. Even if there is only enough food for two wolves, the pack will expect the Alpha pair to eat it all to ensure the survival of the pack.

Gesture eating is my way of demonstrating the owner's primacy at feeding time. It is an element I ask people to apply for a minimum of two weeks. If at all possible, I prefer every family member to participate. By acting as a team, this will allow you to communicate an immense amount of information and will establish all of you at an upper level of the household's hierarchy.

Again, the overriding priority here is to be calm and consistent, and it is essential this is repeated at all meal times during this period. Many people, for practical reasons I can understand, feed their dogs during the evenings only. For maximum impact, I prefer it if dogs are fed twice daily, once in the morning and again in the evening. The technique, below, is simple.

Gesture Eating, Step by Step

Step 1

Bring your dog into the kitchen or feeding area. Before preparing the dog's food, place a biscuit, cracker, or any other small snack on a plate on a raised surface, such as a table or kitchen worktop. If two or more family members are present, each of you should have their own snack.

Step 2

Place your dog's bowl next to the plate. Ensure the dog is watching and prepare its meal in the normal way.

Step 3

Each family member should now eat their snack. This should be done in full view of the dog, but without anyone speaking to or acknowledging its presence. This underlines each person's status as senior members of the household hierarchy in the eyes of the dog.

Step 4

When everyone has finished eating and returned their empty plates to the table or worktop, the dog's bowl should only now be placed on the floor. Again, this must be done quickly, without any fuss. Once the dog has finished the meal, the bowl should be taken away immediately. Again, the message is clear: the leader is the provider of food, and it is the leader that decides when feeding time is over.

It Is Better to Give Than to Remove: "Power Gesture Eating"

A popular argument among traditional trainers is that a powerful way of "showing the dog its place" is taking its food away from it. I could not disagree with this more strongly. It is confrontational and potentially dangerous, particularly when the dog's previous experience is of being denied food. "Power gesture eating" is a means for proving the opposite applies. Nothing is more certain to win a dog's devotion than being seen as a giver rather than a taker of food.

I first developed the technique to help owners who have problems with dogs that become nervous or aggressive at meal times. It is a problem particularly common to rescue dogs, which can snarl and snap at anyone who comes near them while they are eating. It is easy to see why they behave this way. Often these dogs arrive at sanctuaries or rescue centers having been starved close to death. When they are given food, they are desperate to hang on to it. And anyone approaching them is seen as a threat that might take that food away. Instinctively, the dog feels that it must challenge the person with the food.

One of the central aims of my approach is to instill in the dog the feeling that it always wants to be at its owner's side. There is no better way of getting this idea across. The idea is that by following the process above, the dog will form a powerful, positive association between food and its leader. It will see you as a giver. And the dog will begin to realize that wherever it finds you, so too it will find food. The process is described in the panel on the next page.

Packs: Variations on Gesture Eating

Feeding time presents a perfect opportunity for you to demonstrate leadership to the pack as a whole. It enables you to establish in the clearest language that you are responsible for the allocation of food. And it allows you to do so in full view of the gathered pack. Bearing this in mind, it is, however, vital that owners of multiple packs are once again aware of the subtle dynamics at work within their family of dogs. From the outset, each dog should be allocated its own spot, a clearly defined piece of territory within the dining area. This should remain the same at every feeding time from then on. Allocation of these areas should be given a great deal of thought. If there are two dogs that do not get along, it makes sense that they should not have their bowls placed adjacent to each other. You should also try to reflect the alliances within the pack, placing friendly dogs near each other.

The essential gesture-eating technique for packs remains the same as for a single dog. Another thing to bear in mind, however, is that you may easily disrupt the balance within the pack's hierarchy by feeding dogs in a specific order of preference; this could easily create tension within the pack. Given that gesture eating is going to continue for only a few weeks, it would be helpful if an extra pair of hands was available at main meal times. This way, the bowls could be

Power Gesture Eating, Step by Step

Step 1
Set the preparation of the meal up in the same way as normal gesture eating. Instead of putting all the dog's food in one bowl, however, divide it up into three or four smaller portions in separate bowls.

Step 2
As normal, eat a small snack in full view of the dog. Then place its first bowl of food on the floor and step back.

Step 3
As the dog completes the first bowl, place a second bowl of food down in a different area of the room. Stand by the bowl until the dog comes to it, then step away again.

Step 4
Keep repeating this for the remainder of the bowls, making sure each time the food is put down in a different location and that the dog sees who has placed it there.

distributed at the same moment, so as to avoid any hint of favoritism.

If, for whatever reason, you cannot feed the dogs at the same time, you should operate a strict rota system. If dog A is fed ahead of dogs B and C in that order for meal one, then, when meal two arrives, dog A should be fed last, with B and C getting priority in that order. Meal three should begin with dog C being fed ahead of A then B, and so on.

In my experience, it is rare that people encounter problems with gesture eating or the use of food rewards. Owners, in general, find the principles straightforward and easy to apply.

Frequently Asked Questions

My Puppy Has a Voracious Appetite. Should I Gesture Eat Every Time I Feed Him?

In a word, no. Puppies' eating habits differ wildly from those of older dogs. On average, an eight-week-old puppy needs four feeds a day. Experience has taught me that you do not necessarily need to gesture eat each time you feed a puppy; once a day is sufficient to impart the message that you are the food provider. The only thing I stress is that you must not make the mistake of leaving food down all the time. Tempting though this is with a ravenous young animal around, it undermines and negates the message you are working so hard to impart in every other area of the dog's daily life.

Should I Involve Children in Gesture Eating?

Of all the elements of my method, this is the one that lends itself most naturally to involving children. It can easily be presented to them as a game, and it can be great fun to go through this process together as a family. The only word of caution I would issue is that the gesture eating should not become a form of teasing or, even worse, humiliation. It should be done in a matter-of-fact way.

What if My Dog Likes to Snack?

One of the easiest mistakes people can make is to leave food down for a dog at all times. It sends out the worst possible signal: it allows the dog to dictate when meal time occurs. And this simply mustn't happen. For the dog to accept you as leader, it is vital that you retain complete control of feeding time. You should decide when it begins and when it ends, and this must be adhered to rigidly.

If your dog has a habit of walking away and taking breaks during its meal, this should be discouraged by the removal of the bowl. Again, this builds on instinctive pack behavior in the wild. If a wolf moves away from its kill, it is signaling that it is satisfied and the next member of the pack's pecking order can move in. It is not a decision that can be reversed. It is the same here: the dog must learn that, when it steps away from its food, it signals the end of the allocated feeding time. It must learn the consequences of its action. You should not worry about depriving your dog of food; it is not, I can promise you, going to starve. Dogs pick up all the signals within my method quickly, but any message relating to food is picked up that little bit quicker.

Is There a Danger of My Dog Eating Too Much?

Yes, and this is something that all owners must guard against. Once your dog has come around to your way of thinking, rewards should only be used in rare instances. During the course of the early phases of training, however, tidbits will be used on a regular basis, so it is important to ensure that the individual food rewards handed out are both modest and appropriate. I do not want to be responsible for creating a world of overweight, and therefore unhealthy, dogs. By "modest" amounts, I mean that the food handed out should be small in size, say a half-inch piece of meat strip or a sugar-cube-sized piece of cheese. By appropriate, I mean that the food should be part of the dog's natural, healthy diet. I have come across owners who have asked me whether they can reward their dogs with chocolate or cookies. I dread to think what the effect of this would be!

There is another reason why you must exercise caution in this area. The aim of my method is to relieve the dog of

The pack. All my dogs walk without leads, looking to their leader.

Greeting the Alpha. My dogs respect my body space while their cocked-back ears show how contented they are.

Ear positions display a dog's moods

Sasha is relaxed, with her ears pricked, showing she is alert and happy.

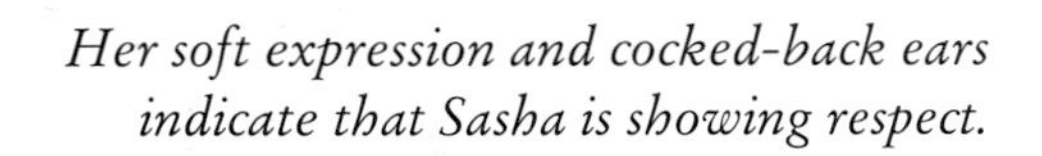

Her soft expression and cocked-back ears indicate that Sasha is showing respect.

Sasha's body is low, her ears down and her eyes anxious, as she demonstrates worry and anxiety.

The classic play pose; Sasha offers to play with Spike Milligan.

Telly addicts: two of my German shepherds being entertained.

Surplus to requirements: Derek with the rubber gloves his owner no longer needs.

Inseparable: Ernest with Gypsy and Kerry.

A reformed character: Dylan, the once delinquent Akita.

Returning to the Alpha: my dogs demonstrating the most vital command of all.

On air: participating in a BBC radio phone-in.

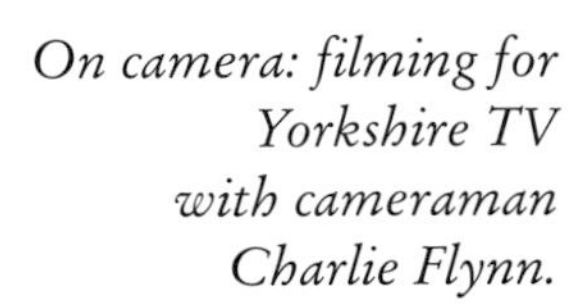

On camera: filming for Yorkshire TV with cameraman Charlie Flynn.

Another satisfied customer — me with Rachel Keys, whose two rescue dogs, Keeny and Peggy, were transformed by my method in a demonstration for the Daily Mail.

My mentor: Monty Roberts and I share a moment after I spoke at his demonstration in September 1998.

> the pressures of leadership, to lift the yoke off its shoulders if you like. The advent of this new, stress-free existence can also have the effect of making your dog put on weight. The reason for this is simple: faced with the role of leading the pack day in, day out, a dog burns off calories. When this pressure leaves its life, it becomes happier, more relaxed—and potentially fatter. There are two easy ways to deal with this, of course. Your dog's diet should be monitored, and reduced if necessary, and the dog should be given as much appropriate exercise as possible. And, unlike humans, it can do this without state of-the-art running shoes or membership of an expensive gym.

More important still, dogs are extremely quick on the uptake in situations where food is involved. It was not merely its rarity value that made the case of Dexter the boxer so memorable, however. It was the fact that it also made me howl with laughter.

Dexter's owner, Tom, had read *The 30-Day Training Guide*, and had applied the method with great success. He and his family had been working at it for three to four weeks and were delighted with the transformation. As Tom put it to me, "That book could have been written about Dexter." He had every bad trait going, from jumping up to pulling on the lead, from charging at strangers to barking at the boundary fence. All of these had been ironed out. The only situation in which he now barked was at times of perceived danger. But here, too, he had improved enormously.

One weekend, however, Tom couldn't help wondering why Dexter had been making woofing noises all morning. Saturdays were normally a quiet time in the neighborhood, yet it was as if strangers were constantly coming near the front door or passing close by the boundary fence. When Tom took a look out of the window, he saw something hilarious. Dexter would run to the garden fence. He would then turn to the house and let out a short burst of barks. No sooner had he done so than a family member would appear with a piece of food reward, as thanks for the warning. The problem was that there were no threats at all. Watching from upstairs, Tom saw that the street was quiet. It has always been well known that boxers are highly intelligent. Even by the standards of his breed, Dexter was a very bright boxer indeed: he was inventing passers-by so that he could be rewarded with food.

When I heard this story, I quickly realized what had happened. In accordance with the method, Tom and his family had introduced the idea of thanking Dexter for his warning of a perceived danger. This is something we will cover in due course. The family had underlined the message by handing out food rewards. It was wonderful to see a family involved in this way; the only problem was that they had failed to ease off the rewards, and the dog had devised a clever way of capitalizing on this.

The lesson here, then, is that food rewards must be reduced and replaced by verbal praise. Certainly by the third and fourth weeks, it should only be used when your dog has behaved exceptionally well in dealing with difficult or totally unfamiliar circumstances. Otherwise you will be conned in the same way as Dexter's owners were . . .

The Power of Playtime

The first forty-eight hours present a real challenge to both dog and owner. It is a time for concentration and discipline. But it is also a time when fun has its place too. The act of play is an enormously powerful tool for building on the early successes of Amichien Bonding. This is a good time to bring this into your armory too. We can see how hugely important play is when we again look at the model of the wolf pack. Wolves do not go out for a casual walk in the way humans do. They only venture out from the security of the den on two occasions: the hunt and when the den and the pack as a whole are on the move, relocating their territory.

A wolf's major source of exercise and social activity comes through play. This is something that every member of the pack takes part in. Even the Alphas enjoy a bit of rough-and-tumble—and they, once more, use it as a means of subtly re-emphasizing their leadership. By ensuring they control playtime, the Alphas remain the decision makers. They do so by retaining control of the whole situation, choosing the timing of play and exercising ownership of whatever objects might be used for the purposes of play. But, more subtly, by creating an atmosphere in which the entire pack enjoys itself, they are providing their pack members with yet another reason to value and appreciate the society they form together.

It is akin to the sort of bonding sessions so popular among large corporations these days. The idea there is that people get together outside of the work environment and have fun. The hierarchical order is not broken—the boss remains the boss—yet the staff's perception of the corporation is once more reinforced. They see it as a community that offers them enjoyment as well as security. Why would they want to undermine or move away from this?

Puppies, in particular, learn an incredible amount through play. It is by tumbling around and chewing and biting on playthings that they develop their athletic abilities and cut their teeth. Yet it is precisely because play holds such a potent power to pass on messages that it must, from the outset, be conducted in the right way. It is another occasion on which you must establish, and maintain, leadership.

I introduced the concept of "toys not trophies" in my first book—it is worth reiterating again here, nevertheless. Again, it is an area best understood by considering it from the dog's point of view. As far as we are concerned, our everyday toys—balls, rag dolls, bones, and throwing sticks—are nothing more than playthings. Yet, if we look at it from the dog's perspective, these objects take on a much greater significance. Within the natural environment, playtime is another crucial opportunity for the Alpha pair to underline their leadership status.

When we look at it from this angle, we immediately see why it is that dogs attach such importance to play. Far from being unimportant playthings, toys are trophies, badges of honor if you will, to be won and lost within the pack environment. It is something that becomes blindingly obvious when we see it in these terms. We have all seen the way packs of puppies wrestle around for toys. The triumphant dog struts around afterward, like a heavyweight boxer with his vanquished opponent lying at his feet. I have also seen situations where dogs become agitated and aggressive because their owners will not play with them, when they are presented with a toy. People tend to dismiss this as playfulness, over-exuberance, or childlike tantrum throwing. In fact, it is a control issue. Dogs believe they are in charge of playtime and behave accordingly. By allowing our pets to believe they have control of playtime and the all-important status symbols that are their toys, we are also allowing them to delude themselves about their place within the pack. This is why it is so important that you impose yourself as leader through play by applying the few simple principles below.

Take Control of the Toy Box

The most straightforward way of imposing leadership at playtime is by assuming control of the toys themselves. Now I am not, in any way, advocating that playtime is eliminated unless you are present—that would be silly and plain wrong. Therefore, I ask owners to leave one or two favorite toys around the house so that the dog can choose to play on its own. At the same time, I insist owners keep the toys used for interaction in a place where the dog cannot get access to them. By doing this, you make sure the potent power of playtime lies in your hands from the very beginning. It is you and you alone who decides when playtime takes place and which toys are used. As for the choice of toys, that is entirely a matter of personal choice. My only piece of advice here is to avoid small toys. Dogs have choked, and in some cases died, because owners have given them objects small enough to get trapped in their windpipes. You wouldn't leave a child to play with something that could choke it, so why would you think of doing any different with a dog?

Don't Get into Contests

As I have explained, to the dog play is never simply for play's sake. It is a contest that it wants to win, and this is an attitude that you should adopt too. For this reason, you should never get into tugging contests with your dog; by doing so, you are allowing it to dictate the rules of the game. At the same time, there is the danger of showing a dog—particularly a larger breed—that it has physical superiority over you. Needless to say, if a dog starts to believe it is stronger than its leader, it will begin to reassess whether that person should retain their status.

The same applies with play biting. Many owners have fun letting their puppies gnaw away at a finger, despite the nasty little nip the pup's young, needle-sharp teeth can inflict. If the biting is allowed to continue, however, you risk getting hurt and, once again, your dog stands to gain an inflated opinion of its status within the pack. This simply must not happen. If your puppy gets carried away and bites too hard, you must immediately remove your hand and end the playtime. The puppy must learn the consequences of this action.

Make Play Constructive

Playing with your dog is a magical part of the relationship, something to be cherished and enjoyed at all times. Given that my method is based around non-aggression, it is often the perfect time to transmit information to the dog. I frequently use play to practice and top up skills like the recall and coming to heel, which we shall come to later. Both of these skills need regular refreshing. I do this simply by moving away from my dogs when they retrieve and return to me with a ball. I then encourage them to come to me. They want the game to continue. They know that, for it to do so, the ball must be back in my hand; the dog wants to carry on playing so it behaves in a way that ensures this happens. Again, the simplest of signals can be the most powerful.

PART THREE:

Days 3–7

Why There's No Place Like Home

The first forty-eight hours are tough, I make no bones about that. But, by the third day, the benefits of all your hard work should begin to shine through. During the remainder of the first week, you can expect to make huge improvements in your ability to communicate with and control your dog. You should, for instance, be ready to develop your first set of basic control skills. You should also be able to move on to the third element of the method, dealing with times of perceived danger such as the introduction of unfamiliar faces and noises. Best of all, you should be able to have some real fun with your dog.

It will be an exciting and rewarding time for all owners. But it will also be a time when you must remain disciplined. To ensure this, I insist that this next phase is conducted within the confines of the home environment. If I have a single recurring message here, it is summed up best in the words of Dorothy in *The Wizard of Oz*: there really is no place like home. There are several reasons for this. Calmness and consistency are central to my method. In the frenzied world we live in, most people regard home as the one haven of sanity and peace where they can rely on finding these qualities. The upshot of this, of course, is that you are relaxed within this atmosphere. As I have explained, relaxation—keeping your head—is a key component to leadership as well. Home is the best place for this.

Home is the most potent place for your dog to learn too. Again, the parallel with the pack applies. It is only when the Alpha pair have led their charges safely back within their own marked ground that the mini society that is the pack can operate on every level. Outside this environment, life is defined by vigilance against attack and the rules of the hunt. At home the Alpha pair and their pack can eat, sleep, play, and socialize more fully. It is here that the young learn the subtlest yet most powerful lessons about the way they are expected to live.

This stage is without doubt one of the hardest parts of my method for most owners. And I understand why. Their natural instincts, quite rightly, are to take their dog out into the wider world—to exercise. Some people say to me, "It's unfair to the dog"; some even say it's cruel. My response is always the same: it is far more unfair and cruel to take a dog out into the world where neither the owner nor dog is capable of dealing with the dangers they are going to face there. I often compare learning my method with learning to drive a car. No one is allowed to take the wheel of a car until they are fully familiar with its controls and know how to operate them properly. No one in their right mind would go out on the road having learned to operate the accelerator and clutch, but having no idea where the brake is! It is no different with a dog. Anyone thinking of taking their dog out into the world before learning this is, to my mind, jumping in a car without the slightest idea how they are going to control, steer, or, most importantly, stop it. Lessons learned at home stay with a dog. Lessons learned away from home fade from the memory.

* * *

If I ever want to illustrate this point, I use the example of the troublesome pet I refer to as "Dog X"! There have been times in the last few years when I've wondered whether I've walked accidentally into the plot of a John Le Carré spy novel. I've had to learn undercover skills that wouldn't go amiss within MI5. For reasons I can understand, some people do not like to admit publicly they have problems with their dogs. None of us is immune to criticism, and for many people their relationship with their dog is an intensely sensitive area. I usually encounter this sort of sensitivity when the job involves someone connected with the dog fraternity—and so it proved once more in this particular instance.

The owner's anxiety was made plain when I went to visit her and her handsome, tri-colored Border collie at her home. As we talked at our initial meeting she actually said to me, "If you see me in the streets please don't acknowledge me, walk on by." It turned out that the lady was a very successful competitor within the obedience-class world. This is a very competitive world. Owners travel the country and take pride in their performances. She had had success with other dogs and was a judge herself. People often ask me how I can square working with people who adhere to these ideas with my disdain for the very word "obedience." My answer is always the same. If there is a chance I can convert an owner to my compassionate method, it is worth taking.

The dog I was asked to look at was potentially one of the best the owner had ever worked with. But he had one flaw in the eyes of the judges. Obedience-class competitions are not dissimilar to the equestrian dressage contests. Owners are judged on their ability to maneuver their dogs according to precise instructions. This particular dog would stop when his owner said and he would turn with her. But, throughout the heel-work part of the routine, he would refuse to walk in the optimum position, staying two or three inches in front of her instead. He would also lean into her, again something that was losing marks. It was very frustrating for her; the dog was getting scored very highly in every area, but when it came to this one area he lost points heavily. Judges would comment along the lines of "If only it didn't lead in front."

The owner couldn't understand why the dog had progressed to this point yet couldn't make that final step. It seemed such a small adjustment to make. To me it was pretty obvious why the dog was holding back. The lady was taking him to weekly obedience classes, held in halls and fields in the village nearby. The classes were using all the old-fashioned ideas: enforcement, choke chains, and commands. My argument, of course, is that force is not going to change anything. That aside, the main reason the owner was failing was that she had not established leadership in the dog's eyes, and, crucially, she was not working in the right environment, the home. Again, the comparison with children is a good one. If they have been taught to behave in a certain way at home, they are unlikely to behave differently in someone else's home. Equally, if they have been allowed to run amok at home, the chances are they will do something similar when they are outside the home.

When I visited the lady at home, it was clear that Dog X believed he was the boss. He was hyperactive and charged around furiously when I arrived. The owner told me that sometimes during the walk, he would run off on his own and she would not be able to recall him. I explained the problem to her. While she was trying to get him to do things exactly as she wanted, the dog instinctively wanted to lead her. In the competition ring he gave every sign of being under her control. Yet the leaning and the walking a few inches in front represented his sole gestures of defiance. They were the dog's way of saying: "Actually, I'm the one in charge here."

I quickly established a rapport with the dog, then I took him into the garden after about an hour. I worked with the lady, using her commands. She was pleased with the results and, as I left, I felt confident she would get the improvement she was after in the competitive environment. I told her to keep her dog away from classes for a couple of weeks and to work on removing leadership responsibility from him in the meantime. She was clearly dedicated to improving the dog and set about her task.

Unsurprisingly, I did not hear from her again for a long time. It was only after the publication of my first book that I came across her—at a book signing. She came up to me smiling warmly yet still pretending she didn't know me. I happily went along with her charade. I was keen to hear how she had been doing and was delighted to hear that her dog had been doing fantastically well in his obedience competitions and had even competed at Cruft's. He had, as she had always hoped, turned into the best dog she had ever had. As she was leaving, the lady touched me on the arm and offered a last, mildly embarrassed smile. "Forgive me, but I took all the credit," she said.

Perceived Danger

Nothing is more unsettling than the sight of something unfamiliar in familiar surroundings; it is the principle that gave Alfred Hitchcock a career. And it applies to dogs as much as it does to us, their human companions. The difference with dogs is that, as the assumed leaders of the domestic pack, they believe it is their responsibility to deal with such threats. Understandably, it is these situations that are often responsible for the most extreme behavior.

Again it helps to take a dog's-eye view of this. Seen from the dog's perspective, even the most trivial domestic event takes on a sinister new perspective. Take the arrival of the morning mail: How would any of us react if we heard a strange noise preceded by the sight of strange objects being thrust through the entrance to our den? Or the sudden arrival of a plumber: How would we feel if we discovered a man we had never seen before walking into our house brandishing a collection of dangerous-looking objects? Precisely—we'd be petrified for our own safety. And if we believed it was our responsibility to guard the den and its occupants, we would be even more distressed.

The domestic home is a minefield of sights and sounds that could easily be construed as being potentially dangerous. Outside, there are the sounds of cars, trucks, airplanes and passers-by. Inside there are the assorted sounds of daily life, from washing machines and telephones to the bumps and bangs of young children crashing around in their rooms. The problem is that humans do not understand this. So, when they hear a dog barking at the sound of a ringing doorbell, they chastise it for being too noisy. When they see it lurching at a postman, they condemn it for its aggression. They fail to understand that, in reality, the dog is dutifully carrying out the leadership role for which it believes it has been elected. Is it any wonder when it proves so resistant to calls to "Stop it" or "Behave"? The dog is utterly confused that, rather than being congratulated, it is being abused by pack members who seem to have joined in the general panic. This is why you must learn to relieve your dog of this responsibility as early as possible within the process.

The true test of a good human leader lies in your ability to deal with times like these. The manner in which you deal with danger is of no less importance in the canine world. By taking charge of a situation, you can take a huge stride toward underlining your long-term suitability to lead the pack. And to achieve this, you must now build on the work you have already done in establishing yourself as the pack leader.

Dogs assimilate easily into domestic environments and—apart from some exceptions—learn to live with many everyday occurrences. Yet some dogs are sent into a blind panic by certain noises. The key to working in all these situations is, again, calmness, consistency, and providing a convincing model of leadership. The dog must first be reminded of its status as a subordinate. It must then be reminded that this role does not require it to deal with the situation at hand.

The best way to illustrate this is by looking at the most common threat to the domestic pack: when outsiders enter the "den." Viewing the situation from the dog's perspective, it is easy to see why it gets so agitated. At the top of its job description is its responsibility to guard the den and all its members from invasion and attack. When it hears a knock on the door and then sees a stranger about to enter the house, the only logical conclusion, to the dog at least, is that an assault on the den is about to take place. Its territorial space is being invaded and it needs to do something to reclaim it. When we look at it in this way, it is little wonder that dogs bark, rush at, or even attack people when they arrive in the home.

How, then, should you deal with this potentially dangerous situation? The first time a visitor comes to your house, you must instantly establish the precedent that it is you who deals with the situation. If your dog simply barks or growls, you should step forward to greet the visitor and, at the same time, thank the dog for its help. You must make sure you do this calmly and quickly. You must never make something out of nothing and dramatize the situation. The effect here is twofold: by remaining calm, you are once more displaying all the credentials of a leader. At the same time, you are acknowledging the dog's help. Remember, even after its demotion from the role of leader, the dog's instincts are geared toward integrating itself into a happy and successful domestic pack. It wants to feel useful, that it is contributing. By listening to your dog, then telling it that it has contributed to defending the den, you are reassuring the dog about its importance as a pack member and your dog will feel good about this.

Of course, all dogs have different personalities. And some may react to perceived danger in more aggressive ways. If your dog does have a history or tendency toward being aggressive, it should be fitted with a collar during the early days, so that you are prepared for every eventuality. If your dog does leap up at a visitor, you must act decisively. The dog should be led away from the visitor by its collar. You should then guide the visitor into the house, and away from the dog. If your dog's behavior is so aggressive that it fails to remain out of the way, you must remove it from the scene.

There is nothing for you to fear at this stage. It is important to remember that my method is a holistic approach. By the time you are fully conversant with it, you will be blitzing your dog with powerful signals that relieve it of its leadership. By the time your first visitor comes to the home, you will hopefully already have begun establishing a new hierarchical order and your dog should be more receptive to requests.

Of course, it is perfectly possible that a visitor may come to the house before you have had time to apply any other element of the method. No matter: your reaction to your dog should be the same, and it should be thanked for its help. The process will have been given a useful kick-start. If your dog does react to the arrival of a visitor, it is advisable that you pre-empt the same thing happening again on departure. The easiest way to do this is by removing your dog to another room or to the garden shortly before the visitor leaves. When you have the home to yourself again, you can bring the dog back in and go through the normal reuniting process. The beauty of this is that it will enable you to tackle immediately whatever short-term effect the stranger's arrival may have had.

The arrival of strangers is not the only situation in which a dog feels the pack is under threat. When a member of your family leaves the home, your dog feels equally challenged as leader. Again it is easy to see why this is when we think about it: the dog has no concept of time, in the human sense at least. So when it sees a member of the household—of the pack for which it believes it is responsible—heading through the door, it has no idea whether this person is ever going to return. It is no surprise that dogs become so agitated when family members leave the home; they are trying to prevent the pack breaking up forever.

Again, the way to disarm this potentially explosive situation is by underlining the recurring message that this is no longer the dog's responsibility. It no longer has to worry about these matters. The dog should be offered another warm "Thank you" and led away from the exit door. Ideally, this should be done through using a food reward. The dog will make a positive association the next time this occurs. In cases where the dog offers greater resistance or aggression, you must remove it in the same way as earlier, either by the collar or—in extreme cases—by a lead. It must be taken into another room so that the person can leave.

If you own a very nervous dog, you must once again show particular patience. Rather than confronting visitors, such dogs may run away and hide. In extreme cases, they may wet themselves. The crucial thing here is to leave the dog alone. It has chosen the option to take flight, and this option must not be removed from it. As with reuniting after a separation, these dogs must be given time to overcome their nerves. They must be persuaded that human beings are not automatically associated with pain. They must be given time to heal, and time to learn at their own pace.

It is important here to be aware of the direction in which this element of my method is heading. I have seen this process happen so many times now that I have become accustomed to it, but it may seem difficult to deal with at first. I can assure you that, in time, your dog will learn to accept your leadership. The arrival of a stranger will be greeted with a minor reaction. And, in time, a simple "Thank you" will return your dog to its usual relaxed state.

* * *

One of the most extreme cases I've encountered that involved perceived danger was Hamish, a lovely Jack Russell-Border collie cross I met at his owner Sarah's home in a small, northern Scottish village near Banff. Hamish had lived with Sarah, her husband, and their children since he was six months old. Throughout his two years in the house, he had been unbelievably nervous about visitors, passers-by, and, most of all, the sound of the telephone ringing. He would instantly become wild-eyed and start chasing his tail, spinning like a top as he did so.

Sarah was passionate about dogs. She worked as a dog walker for professional people in her area and was involved with the Good Citizen scheme, encouraging owners to teach their dogs to avoid antisocial behavior. She knew how much Hamish was suffering at these times. She could feel his anxiety and pain—and was prepared to do anything to ease it.

In the two years since Hamish's arrival in her home, Sarah had employed a psychologist, a behaviorist, and a specialist "dog trainer." She had also been drowned in well-meaning advice from friends and family. As a consequence, she had received some rather extreme remedies. The behaviorist's big idea was to fill a tin can with pebbles! The theory was that this would be used as a distraction whenever Hamish got agitated. Sarah would either shake it in his face or throw the tin on to the floor near Hamish in an attempt to take his mind off things. It hadn't worked, needless to say. (All it had done, in fact, was give poor Hamish another potential threat to get in a funk about!)

The "dog trainer"—and I use these quotation marks deliberately—suggested Sarah shake Hamish by the scruff of his neck. She hadn't even attempted this. She loved her dog so much, she wouldn't dream of being aggressive or violent toward him. Finally, and perhaps most ludicrously of all, the psychologist had told her to shout at Hamish. Whenever he went into a spin, she was to place her face next to him and bawl him out. Goodness knows what she was supposed to say. Sarah had rejected this advice, too, just as she done with the suggestions that she should smack him with a newspaper and throw water over him; she simply wasn't going to indulge in bullying her dog into submission.

Sarah had read my book but had struggled to impose herself on the situations when Hamish became most agitated, during times of perceived danger. Her plight had somehow come to the attention of the production company behind my television program. The format of the show involved me applying my method to a group of dogs with a variety of problems. It was a hugely enjoyable experience, not least because it introduced me to some of the more interesting canine characters I have come across. I should point out that, as is the rule with the cases I tackle on-screen, I had no prior warning about Hamish or his problems. I went into Sarah's home "cold," as it were. It did not take long to see what the problem was, however. Within moments of my walking in, he became extremely agitated. He began to spin around in a really crazy fashion. Sarah was in tears as she explained that, in addition to this, he had this habit of biting his tail. She said that she thought he was beyond help.

Without speaking, I took hold of Hamish's collar and sat down on a sofa, with Hamish sitting on the floor alongside me. I wanted to just hold him gently there. I exerted no force whatsoever. If there was any pulling, it was coming from him. I could sense how deeply distressed this dog was; his body was absolutely rigid, there was not a fiber in him that wasn't jangling, and his eyes were wild. This was hardly a surprise, given that his living room was filled with myself, a cameraman, a sound engineer, a researcher, and the television show's presenter, Paul Hendy.

I explained to Sarah that I was going to hold Hamish in this position until his body was relaxed. His reward would then come in the form of his release. At the same time, I asked everyone else in the room to remain quiet and still, and, most importantly, to avoid looking at or making eye contact with Hamish. I held him for maybe a minute or so. Eventually, I felt the tension begin to ebb away. When I felt his whole body relax, I let the collar go.

The next step was for Sarah to try this. She did so with good results. Soon after Hamish was released again, one of the crew made a tiny noise. Hamish immediately went into his whirling dervish routine. So I asked Sarah to take hold of him again. "Every time he does this, this is what you must do," I told her. Sarah was very emotional, but she could see we were making some progress. By the fourth time we went through this routine, we made a real breakthrough. Almost as soon as Sarah sat down with Hamish alongside her, he relaxed.

The next step was to set off the phone. By now, I had programmed Sarah's phone number into my mobile. No sooner had I triggered the ringing than Hamish once more freaked out. His behavior this time was even worse than previously. At one point, he was sitting on the floor with the tip of his tail in his mouth, shaking uncontrollably. Again, I took the lead and brought him to sit alongside me. I then triggered the phone again and held him in position until he relaxed.

By the end of my time with Sarah, Hamish was doing well.

Such was his anxiety, however, that I warned Sarah that it might take her months to totally relieve him of the stress. I asked her to strengthen the process when the phone rang by excluding Hamish from the house and leaving him in the garden. When she let him back into the house, I wanted her to go through the reuniting after a separation routine and the five-minute rule so that she could apply a double whammy of signaling.

I left Sarah's house at 4 PM on Tuesday. At 9 AM on Friday morning, she called me at home. Once more she was in tears—this time of joy. "It's stopped," she said. As usual, I urged caution. But the more she told me, the more convinced I was that Hamish had finally been relieved of the burden of guarding the house. Sarah told me Hamish had slept a lot, a telltale sign of a dog that is recharging its batteries after a hugely emotional period. As we were talking on the phone, someone walked past the front window of Sarah's home. Seventy-two hours earlier, this would have sent Hamish into orbit. Sarah told me that, just like the previous evening when the phone rang, he had barely raised his head and had gone straight back to sleep. Days before, I'd warned Sarah that I didn't possess the power to change things like some fairy godmother. I told her she faced a long, hard road. Even I was amazed at how quickly things had changed. "What was that you said about not having a magic wand?" she laughed.

Pushing the Boundaries: Preparing for the Walk

By now, your dog should be comfortable within its new environment. It should be familiar with its sights and sounds, its household members, both human and canine. And, most importantly, it should have begun to understand its new, subordinate status within the hierarchy of its domestic pack.

As it learns to relax, the next natural step will be for you and your dog to push the boundaries of this environment. Your dog will, understandably, want to explore the world more fully; I believe in taking every possible opportunity to harness the dog's positive instincts. This therefore presents the perfect opportunity for you and your dog to move forward by laying the groundwork for those first steps into the wider world—and the day when you take your dog for its first walk.

Again, it is important to stress that all dogs must be treated as individuals and allowed to develop at their own pace. In general, I recommend that you wait seven days before actually venturing out. Some owners, however, will have to wait much longer before venturing out. The reason for this is simply that not all dogs will be capable of progressing this quickly. Rescue dogs with particularly nervous dispositions may have to wait up to a month before they are ready to venture back out into a world that has, after all, treated them appallingly in the past. Puppies, too, should not walk out into the wider world until they are three to four months old at least. I advocate keeping puppies away from any risk of infection until two weeks after their second vaccination—about fourteen weeks into their life. Well-organized puppy playgroups provide a useful means of socializing and providing play during this time.

For most dogs, however, key controls at home can begin from day three onward. Many owners, I know, find it hard to wait this long to go out on a walk. But the key thing to remember here is that, once out in a strange and unfamiliar world, your dog is going to rely on its newly elected leader to guide it every step of the way. To be able to do this, you need to be able to control your dog via the lead. So the next four or five days must be spent introducing, developing, and perfecting the key controls: the sit, heel work, the turn, the wait, the stay, and the recall.

Basic Control 1: The Sit

The sit is the cornerstone of an owner's repertoire of controls. It will be a fundamental part of life, particularly useful when doing things like grooming, administering medicine, or fitting the lead before the walk. It is vital, then, that this primary and most basic technique should begin during the key, first phase of the training.

Step 1

You want your dog to make a positive association with the act of sitting. To do this, take a piece of food, show it to the dog, then bring it toward and over its head. As you do this, slowly say the word "Sit."

Step 2

As the dog follows the food with its eyes, it should arch its neck backward, so that its whole body tips back and it eventually ends up sitting down.

Step 3

The dog should be rewarded with warm praise, maybe some smoothing of the head, and, of course, the right to eat the morsel from your hand. This should be done as soon as the dog's bottom touches the ground, so it knows that the word "Sit," its action, and the ensuing reward are inextricably linked.

* * *

Use this technique sparingly at first. Don't, for instance, ask your dog to sit each time you ask it to come to you. There is a good reason for this: remember, dogs are deeply manipulative creatures, and they also know when they are on to a good thing. Remember, once a dog has grasped a principle, it may

try to manipulate you into repeating the process again and again. I have seen dogs sit there, staring angelically at their owners, expecting a reward each time. It is vital that this does not happen, as the object of the exercise is to underline your status as leader. The dog is not going to grasp this if it is able to influence when food reward is dispensed. Only the leader should decide this.

Basic Control 2: Heel Work

No one wants to go for a walk with a dog that is trying to pull your arms out of their sockets. No one wants to go for a walk with a dog that is panting and being dragged along like a reluctant conscript. It is a misery for everybody concerned; there is no joy in it. So the goal here is to train your dog to walk closely by your side at all times, and to do so happily on the end of a lead. The dog should slow down, speed up, and stop

Frequently Asked Questions

What if My Dog Doesn't Respond?
Dogs are not mind-readers so you cannot expect them to understand what you want immediately. If at first you don't succeed, try again. It may take several attempts to get this right, but you must remain calm.

When I Move the Food over My Dog's Head, He Shuffles Backward on All Fours.
Move to a wall or a closed door. Place your dog with its back to the wall or door. It cannot now move backward and should tip onto its bottom when the process is repeated. Again, success may take time, but your dog will not need teaching a second time.

What if My Dog Still Refuses to Sit?

Gently place a hand behind your dog. The hand should only touch its bottom ever so lightly, in the way a baby is supported before it learns to sit up unaided. Force has no part in my method; the hand is only there as a stop. Pass the food over the dog's head again, repeating until it tips over and sits of its own free will.

in harmony with you. As before, this is something that should be developed in slow but steady stages.

Step 1

The first thing you must do is elect the side you want your dog to walk on. Most people prefer their dog to walk on their left. I know that within the gun-dog fraternity this is something that is encouraged because most people are right-handed and it leaves the right arm free. There is absolutely nothing wrong with training your dog to walk on the right. The key thing is that, once the decision is made, you stick to it. In what follows, I am assuming you want your dog to walk to the left. If you would prefer to have your dog on the right, simply reverse the instructions as appropriate.

Step 2

To begin heel work, you should work in a garden, a corridor, or a room with sufficient space for you to walk a dozen or so paces to begin with. This should also be done without the use of a lead. It is vital at this stage that your dog retains the option of flight; to deny it may result in aggression.

You should first turn your back on your dog, making sure as you do so that you have a piece of food reward ready. Put the food in your left hand and bring it down along the side of your leg until it reaches a height equivalent to the dog's nose level. As you are doing this, you should call the dog's name and request it to "Heel." The presence of the food reward should bring your dog to your side. If it approaches as requested, your dog should be given the reward and praised warmly.

Of course, this is something that may take time to get right. If, for instance, your dog is attracted by the smell of the food and arrives at your side uninvited, then it must be ignored for a few minutes. The routine must then be started again. Similarly, if your dog does not appear, the exercise should be abandoned and

tried again at least an hour later. There should be no interaction between you and your dog during this cut-off period.

Step 3

You must now begin walking slowly. If your dog remains in position at your side for the length of the walk, it should again be rewarded and praised at the end. It is vital to remember again at this point that your dog is not a mind-reader; it is up to you to speak its language. So, if it wanders slightly away from you, you should encourage it to return to the correct position by the use of a positive association. A supply of food reward should be ready again. The dog should be reminded to "Heel" once more. Again, the key thing, as always, is that you remain calm, that your pulse rate remains low. If your dog does not get it right immediately, be patient and keep repeating the exercise.

If your dog becomes agitated or loses interest and starts jumping around, the exercise should be abandoned and again

there should be no interaction for at least an hour. Neither humans nor dogs are capable of digesting important information when they are in an agitated state, so allow things to calm down and start again later.

Step 4

As your dog learns to walk in harmony with you, the length, speed, and direction of the walk should be changed. You should stop and start again occasionally, using a few softly spoken words so as to gain the dog's maximum attention. In the outside world, you and your dog will have to negotiate all sorts of routes, and this is vital preparation for the day when you face this test together for the first time.

Introducing the Lead

An observer watching a dog that has learned to walk to heel in the right way might well imagine it is connected to its owner by an invisible line. The next stage of the process is to add a visible line in the form of a lead. In most respects this is the same as the heel routine above.

Step 1

Bring your dog to heel. As it stands at your side, you should place the lead gently and without fuss over the dog's head.

Step 2

You should now begin walking, encouraging your dog to stay at your side as normal. If the dog begins pulling on the lead, you should stop and calmly stand your ground. Do not get into a tugging match with the dog; the lead is not a weapon to be used to jerk or yank an animal around with. The dog should be asked to once more come to heel. If the groundwork has been done properly, your dog should return to its starting position easily. The walk should then be resumed once more, with you rewarding the dog each time it successfully completes its length without your pulling and calling it back to heel each time it tugs on the lead.

Step 3

You should slowly extend the length of the walk you take, using the available space to its maximum effect. As you develop this skill, you should become less and less aware of the lead even existing.

Basic Control 3: The Turn

It is a very rare walk indeed that heads in a single, straight line all the way. It is therefore going to be vital that you learn to make turns using the lead. These, again, are maneuvers that can be practiced and perfected at home. Again, I am assuming here that most people will walk with their dog to the left. If, however, it is the right, simply reverse the instructions.

Turning Right

As you pivot round, you should lead with your right leg. Leading with the left leg will cause all sorts of problems: the dog will be blocked from turning right and you and your dog run the serious risk of tripping over each other. As you turn, use a distinctive word that your dog will always associate with this maneuver from now on. You are free to use whatever word you like, but the traditional word in the dog world is "Close." The most important thing is that your word has a distinctive sound.

As you turn your body, your dog's head will move with the lead and its body should angle around to the right as well. This is something that should happen instinctively. If you have applied all that has gone before properly, your dog should by now be paying attention to you at all times, and it should want to be with you at all times too. The dog should be doing these things of its own free will.

Turning Left

This is a slightly more complex maneuver, so I advise you to begin by gathering up your lead so that there is no slack. This should ensure that the dog is at your side as you prepare to make the turn. You should extend your left leg out as far as you can ahead of you. At this point, your leg should be touching the dog's neck area. Again, you should choose a word to be associated with this control; the traditional word is "Back," for the simple reason that the dog is being asked to drop back.

As you pivot left, your dog should naturally drop back. And because your body is now gently applying pressure to your dog's body, the dog should turn naturally to the left with you. The key here, as always, is that there is nothing sudden or violent about this. It should happen smoothly, seamlessly, and calmly, with plenty of warm praise when things go well. Soon the dog should be reacting naturally to the movements of your legs.

Basic Control 4: The Wait

The ability to bring a dog to an instant halt is vital—both on and off lead. It is important, then, that a new instruction, the wait, is introduced at this time. It is done simply enough. As you develop heel work, you should introduce sudden stops accompanied by the short, sharp, but non-threatening instruction "Wait." After a brief few moments you should get on your way again, using the instruction "Heel." Once more, there is a strong case for using food reward as a means of getting this control right. This may be one of the most important lessons your dog will learn, and it should carry positive associations from the outset.

Dual Controls

I am certainly not an advocate of gadgetry; my aim is to replicate as much as possible the sort of natural behavior found in the wild. And when did anyone ever see a wolf using a clicker to teach one of its cubs how to behave? In the real world, however, there has to be an element of flexibility and I accept that, for some people, controlling a lively dog in the early stages is going to be a difficult process. In the case of rescue dogs that have a history of bolting after cars, jumping at or attacking people, or simply running off, this is not something that can be left to chance.

For this reason, I advocate that in some instances owners use what I call "dual controls" and, in addition to the collar and lead, they use a head brace. Each is controlled from a separate hand, giving you twin means of managing your dog. If the dog has a habit of suddenly swinging its head around, the head brace can be used to bring it back into position. This is only a temporary

Turning Right: Stages 1–4

Turning Left: Stages 1–4

measure, hopefully. Once the dog has got the message, the head brace should be discarded.

Basic Control 5: The Stay and the Recall

You and your dog will step into a world full of dangers. It is therefore vital that you are able to freeze the dog in its tracks at any hint of danger. Equally important, you must be able to call it back to you, wherever it is. Therefore you must master two new controls here, the stay and the recall. In essence these are extensions of the sit and the come.

Step 1

Ask the dog to come and then sit. By now these should both be instinctive for it. You and your dog should now be facing each other. Take one step back and, as your weight is transferred to your back leg, extend an arm and place the flat of your hand about one foot in front of the dog's face. At the same time, softly say: "Stay."

Step 2

Complete the transfer of weight so that there is now one stride separating you from your dog. If the dog comes to close the gap, your weight should be transferred back onto the front foot. Your dog should again be asked to sit on the original spot. If necessary, gently place your hand on its chest to position it there. Repeat the move of earlier. It will not take the dog long to realize you are not running away and it will stay.

Step 3

When you can stand with both your feet together one step away, you should stand for a few seconds and return to the dog and reward it with praise. By this stage, any food reward should be less necessary.

Step 4

The previous routine should now be practiced with you moving back to a distance of two strides away from the dog. Again, if the dog comes, or even begins to fidget or raise its bottom in advance of a move, you should quickly move back to the dog and repeat the sit instruction.

Step 5

Once you are standing successfully two steps away, your dog should be asked to come. It should be rewarded with food at this point; it has made an important step forward. The sit and come routines have successfully evolved into the stay and recall.

Step 6

This should be extended until your dog stays for thirty seconds and you can turn away from it without the dog moving toward you. If it does make a move, turn back and repeat the stay command.

Grooming

A dog is a source of great pride for most owners. Many spend long hours, not to mention large chunks of their hard-earned income, ensuring their pets look their best. Yet this area can be fraught with problems. I have come across cases where behavioral problems with dogs have prevented owners from grooming them, sometimes for weeks or even months. This, clearly, is a situation that no owner can allow to go untreated.

Once more, it is helpful to view this at first from the dog's perspective. To the dog, like the wolf, decisions regarding where, when, and with whom it does its grooming are closely associated with leadership. Within a wolf pack, the Alpha pair help groom and clean each other. It is not uncommon for subordinate pack members to play a role, too, licking their leaders clean—but only when invited to do so. Given this, then, it is easy to see why dogs can be so sensitive to the idea of their owners running a brush through their hair uninvited.

* * *

The story of a tiny Yorkshire terrier called Bobby illustrates how extreme the problem can be—and how it can be overcome. Bobby lived with his owner, Mrs. Pearce, in a village a short drive from my home. Bobby was a very nervous dog: he was terrified of leaving the house, barked at visitors, and was capable of nipping at anyone who came too close for his comfort.

It was his biting that prompted Mrs. Pearce to call me. But it was during my first visit that I learned the extent of her grooming problems. Mrs. Pearce told me that she could not go anywhere near Bobby with a brush. A Yorkshire terrier's coat grows quite long and needs regular attention. Out of desperation she had been forced to go through a routine whereby, every two months or so, Bobby was sedated by a vet before having his entire coat clipped off by a professional dog groomer. There were, of course, all sorts of reasons why this could not continue, not least the fact that a dog should not be given anaesthesia on such a regular basis.

During my visit, it was clear that Bobby was a very anxious dog. When I arrived, he scampered at me, stopping at my feet and barking for all he was worth. It took a little while to exhaust his repertoire. Once he had calmed down, I made the first step toward easing Mrs. Pearce's problems. After observing the five-minute rule, I asked Bobby to come to me. At the same time, I placed Mrs. Pearce's grooming brush on the floor, about six inches in front of me. Bobby recoiled at first but came back toward me when I called him a second time. It was important at this point that he knew he had the flight option available to him. Eventually, he approached the brush and me.

By this time, I had been working with Bobby for an hour and a half. I knew I was not going to make more progress with the brushing in one day, so I asked Mrs. Pearce to incorporate a plan to introduce grooming into the thirty-day program ahead of her. This was not the first such case I have come across, and nor will it be the last. Rescue dogs in particular can be incredibly wary of anyone laying a hand, let alone a brush, on them. Whenever I work with a dog such as this, I slowly encourage them to approach their owner so that the owner can introduce a brush across their body without them recoiling or snapping. This is what I asked of Mrs. Pearce. Each day, I wanted her to encourage Bobby to take a step closer to the brush. I told her not to rush. As ever, I emphasized the need to first present herself as a convincing and decisive leader; but at all times, too, she had to allow Bobby the option of flight.

When I spoke to Mrs. Pearce ten days into the method, she had progressed to such an extent that she could now hold the brush in her hand when Bobby came to her. At that stage, she did not yet feel confident enough to try stroking him with the brush, however. By the time she had begun the third week, she had crossed this hurdle and had begun to run the brush, very gently, across Bobby's chest.

Of course, this work wasn't going on in isolation. Mrs. Pearce was bombarding Bobby with signals from all the other elements of the method. Nowadays, Bobby can have a haircut without any kind of sedation. And Mrs. Pearce grooms him at home herself. The bond between the two of them has grown deeper than ever—all thanks to the patience and understanding both showed when it was needed.

Frequently Asked Questions

Can I Groom a Pack Together?

My answer to this often-asked question is no. In my experience, it is far better to groom members of a pack of dogs separately. This is partly sentimental in a way: to me, grooming is an enjoyable and intimate part of an owner's relationship with their dog. It builds a bond between the two. But more importantly, by separating your dogs, you avoid the possibility of giving out dangerous signals regarding one dog's status.

This can happen in two ways, firstly by grooming—and thereby seeming to favor—one dog ahead of another. But it can also happen if you raise dogs onto tables for grooming. One only has to look at dogs that jump up to understand the importance of height within the canine mindset. Dogs that jump are, of course, trying to gain equal status with their owner. So it follows that, if a pack is brought into a room to be groomed together, the sight of one dog being raised above the rest transmits a potentially explosive message. It is signaling that this dog is being elevated in status. As a result, when the dog is lowered back to ground level it is almost certain to be challenged by canine peers who are suddenly uncertain of its place within their pack. In my view the best, and simplest, way to avoid this is to keep the dogs separated. It will make for a more enjoyable and trouble-free time all round.

PART FOUR:

Days 8–14

Home and Away

What is about to happen is an extremely significant moment in the relatively short history of the pack in its present hierarchical form. The family is about to step out of its den into the wider world. It is important at this stage to go back to first principles, and to once more imagine this situation from the dog's perspective. For the dog, exiting the den can only mean one thing is happening: it is heading off on the hunt.

Within the wolf pack, this is a time when the Alpha pair, from the outset, stamp their authority on their subordinates. It is the Alpha pair that decide when the hunt occurs; it is the Alpha pair that check to see whether it is safe to leave the den; and it is the Alpha pair that lead the pack into the outside world and choose the direction in which they are to head.

It is vital that, each time the pack goes out on the walk, you take charge of each of these aspects. And it is now that all the groundwork of the previous seven days should pay dividends. By now, you should have clearly demonstrated yourself as leader. And your dog or dogs should have consented to this and agreed to join this newly formed pack. Your dog will have learned it has no reason to fear for itself when you are in charge. It should now be willing to follow you out into the wider world and to face whatever is waiting for it there. All in all then, the big day has finally arrived.

The Walk

Step 1

Call your dog to you, pop on the lead, and get it to walk to heel with you to the door. If the dog gets overexcited or rushes ahead of you, calm things down by holding your ground. If necessary, postpone the whole event if your dog misbehaves; it will quickly get the message.

Step 2

The crucial thing now is that you are the first person to cross the threshold. If your dog tries to move forward and get out ahead of you, you should step back and bring it to heel. As usual, this should be done calmly, without any histrionics or raised voices. You should then head off again, once more leading the way through the door.

Step 3

Once outside, your dog should walk to heel. Don't worry too much if it gets a step or two ahead; you should only act if there is any tension on the lead, again calling the dog to heel. Each time your dog returns to its proper position, reward it with quiet, sincere praise.

Step 4

The next crucial moment comes when you and your dog reach the boundary of your home. It is imperative here that you reinforce your leadership by choosing the direction the walk takes. So, if your dog begins heading off in one direction, execute a smart about turn and start walking the other way. If your dog veers off again, you should turn once more. Repeat this until the dog understands that it is not in charge of the walk. Again, the key thing here is that no words are spoken. Your dog must be made to voluntarily think: "Where are we going?" It should then come to the conclusion that it is not up to it to decide, and it should follow you, its leader, of its own free will.

The first walk should be a gentle introduction for both you and the dog, and it should continue this way for the next few days. You should build up your confidence with the controls you have developed so far. Your dog should become used to the idea that, when it steps out into the world with you, it always returns safely home. At this point, I do not recommend you let your dog off its lead. The consequences of rushing into this could be dire: the dog might panic, bolt, and do itself terrible damage. Patience, as ever, is a virtue. Apart from anything else, both you and your dog should take the time to enjoy this wonderful experience. There is immense joy to be had in leading a dog that is light on its feet, wagging its tail, and happily exploring the exciting new world that is opening up around it.

It is also important for both you and your dog to enjoy yourselves. The whole point of my method is that you both learn to live with each other in a happy, pressure-free way. Every walk does not have to be a route march; you should let your dogs enjoy the scenery too. I would also caution people with dogs under a year old to be careful not to overexercise them. Walking, particularly on hard surfaces like roads or pavements, can put pressure on growing bones and joints; most agility classes do not accept dogs under the age of eighteen months, as dogs are still growing at this age. By this stage, both you and your dog have come a long way, and you should both take some time to stop and smell the roses.

Pack-Walking

A good owner should never take their leadership of the pack for granted. Even after successfully establishing yourself as leader, you must be aware that this status is going to be challenged on a regular basis. Just like any good chief executive of a company, you must be aware of the individuals you manage, and should know the forces that operate between them. If you own a pack containing two or more dogs, you must always remain aware of the dynamics within that pack. If, for instance, there are males and females, you must be ready for the females to come into season. This is a time when the dog's instincts are at its most highly tuned; along with survival, reproduction is the most important part of a dog's life.

Similarly, as your dogs grow older, you must be aware of the evershifting changes within the pack. I am often asked to deal with situations where a young dog is taking control from an older dog. My honest reply in these cases is that there is little I can do; a dog's nature has to be allowed to express itself. But, provided the dog believes its owner is leader, the changes will be assimilated without any harm coming to anyone. Just as parents can never choose their children's friends, so it is beyond you to control the alliances and friendships that form within your pack. All you can do is temper the behavior within that pack by exerting strong, clear, and compassionate leadership. And that leadership will never be more necessary than when the pack first go out for a walk together.

It is worth stating here that walking a large pack is a very skilled and possibly dangerous thing to do. If there are strong rivalries and personalities within the pack, it is easy to inadvertently engineer a situation where a fight breaks out. If you fail to deal with this situation, the pack will have succeeded in exposing a frailty in your leadership, and all the hard work that has gone into establishing leadership will have been for naught. For this reason, I do not recommend that you take out a pack of more than four dogs unless you are highly skilled and experienced in looking after them.

Preparing a Pack for the Walk

Given the potential pitfalls of taking a pack out en masse, the build-up to this point must be gradual. You must begin the process with a series of one-on-one sessions with the pack members. Each should be taken out on a short walk to develop its individual relationship with you. As I have mentioned before, you must be extra careful when returning to the home with each dog. This will be the first moment two dogs have reunited after a separation since the new method was introduced, and it is highly unusual for the pack to be split up. The dog or dogs that have remained at home will go into a repertoire with the dog that has been outside. This can be a time of aggression, so I recommend the dogs are reunited one at a time—always ensuring there is never one dog on its own.

When each of the dogs has been out and has responded well to you, preparations can go ahead for the first pack walk. This will be a vital test of your leadership. It is a time to exhibit the strong, silent, and decisive body language that your dogs now expect. This will be particularly important as the leads are being placed around your dogs' necks. It is in a dog's nature to get extremely agitated and excited at this point. It regards this, after all, as preparing to go on a hunt, and therefore needs to get its adrenaline pumping. Your job is to remain calm and to provide the cool, decisive leadership expected of you.

There should be no move toward the door until the entire pack has calmed down. If your dogs do not do this, you should exercise the ultimate sanction by removing the leads and thus signaling to the dogs that the walk has been canceled. Provided your dogs do calm down, you should head off to the door, ensuring that—as with a solo dog—you, the leader, are the first across the threshold.

Pack Controls

In almost all respects, the preparation and execution of the walk is the same regardless of whether you have one or more dogs. There is, however, one extra control that will be required: the group heel. Obviously, it is simply not possible for every member of the pack to walk tightly to your heel—not without both you and your dogs tripping over each other, that is! You therefore need to develop a request that brings all the dogs into a tight formation at your side. I personally use the phrase "Come in" or, sometimes, "Together." Again, this is something that can be practiced at home, prior to the first walk. You should immediately reward those dogs that come to your side correctly, both as a positive association for them and as a negative one for those who have not responded.

The positions your dogs adopt will vary each time the walk is underway. The key thing is that they come together immediately when you issue the instruction. If one dog is spoiling the walk for the rest, you should take the miscreant home, release him from his lead, and leave him there. The remaining dogs should then set off with you again. The dog that has been banished will quickly get the message.

* * *

Of course, some dogs—particularly rescue dogs—may be unable to enjoy a walk due to some past trauma in their lives. One such dog, Spike the aquaphobic walker, came to my attention through my television program. Spike lived with his owners—Jo, Paul, and their daughter, Katie—in North Yorkshire. They had taken him in from a rescue center. Like so many rescue dogs, he had a range of problems. What made Spike unique, however, was his dislike of crossing over water. The family lived near the coast and loved walking on the pier. They also lived near a river and several bridges. Spike would not cross any of them.

The moment I met him, it was clear to me that Spike was a dog utterly stressed at being the leader of his pack; his eyes were bulging out and, during my initial spell with him, he tried to leap all over me. He was a strong dog and I had to hold him in position at one point.

An owner is never going to understand all the elements of their dog's psychology. Where the roots of Spike's phobia lay, it was impossible to tell. I have come across dogs that have been so traumatized by certain events that they could not face any situation that reminded them of their past. Previously, for instance, I wrote about a dog that had been found at the side of a motorway, having seemingly been thrown out of a speeding car. It was little wonder he had a phobia about traveling in cars. Perhaps, then, Spike had suffered a waterside experience that had scarred him mentally in the same way. Perhaps he had been thrown into a river? Perhaps he associated bridges with this? I was not going to be able to answer these questions. What I did know, however, was that, by releasing Spike from the responsibility of leadership, I would be able to help him overcome this fear.

I left the family to put the principles of Amichien Bonding into practice. Spike proved a fairly tough nut to crack. At one point, when Jo called him to her, Spike not only ignored her request but looked her in the eyes and deliberately urinated on his blanket. Jo called me in a bemused state about this. I explained that this was his way of challenging her, of testing her authority. I told her to take the blanket away, and to leave him to sleep on the bed without it that night. He had to be taught the consequences of his actions.

By the end of the second week, Spike had begun responding well. Jo and Paul started to take him out for short walks. They lived in a small, picturesque village with a pond, and Spike's phobia was such that he couldn't bear to go close to the pond. Now, I asked his owners to start walking near the water, while always making sure that they remained between Spike and the pond as they did so.

The first day, Jo and Paul walked a fair distance away from the pond. As it loomed into view, Spike glanced apprehensively toward the water. Jo rewarded him with a tidbit and some warm praise and continued walking. Over the next four days, they took the walk ever closer to the edge of the pond. Spike was wary but went along with them; the confidence in their leadership built up in the home was paying off. Day by day, Jo and Paul were demonstrating to Spike that he was always one hundred percent safe at their sides. After a few days, Spike was walking on the pond side of Jo and Paul. The first breakthrough in his water phobia had appeared.

Spike's biggest test still remained. Twenty-three days after his owners had begun the method, they came to the moment of truth: the crossing of a bridge. The bridge they chose was near their home. As if there was not enough pressure on Jo, Paul, and Katie, the television cameras were there to record the event. As if this challenge was not great enough, there had been torrential rain in the days before, and, as a result, the river's levels were extremely high and the water was traveling far faster than normal.

Jo and Paul were very apprehensive as they walked up the ramp leading to the bridge itself. As the cameras rolled, they reached the beginning of the bridge, but would Spike agree to cross it with them? For a second or so, the signs were not good. Spike put his first paw on the bridge and hesitated. Fortunately, Jo was alert and led him on immediately. Spike looked up at his owner for reassurance and moved on. That was his last moment of hesitation; to everyone's delight, Spike strolled across the bridge without any trouble at all. He looked totally unconcerned. The relief all around was immense—not least for me.

The program director was eager to get a variety of shots of Spike's moment of glory and asked Paul and Jo to try leading him across again. Spike trotted across the bridge as before, and kept on doing so again and again for the benefit of the cameras. At one point, he even stopped in the middle of the bridge with his owners and stood there, staring at the raging waters below him. Talk about a transformation! By the end of the afternoon, Jo and Paul were sick of the sight of the bridge. And happily exhausted with it.

PART FIVE:

Days 15–21

Off the Leash

Owners often ask me what their goal should be—what, in an ideal world, they should expect from their dog. My answer to this is always the same. All an owner should want from their dog is, firstly, that it stays with them of its own free will and, secondly, that the dog should do as they ask but, again, of its own free will. This is the point you should now have reached through following my method. If this is the case, the beginning of the third week will represent a significant landmark for both you and your dog. The quiet but nevertheless dramatic revolution that saw your dog deposed as leader is behind you. Now you can look forward to developing a deeper and more enjoyable life together. More work lies ahead of you, but it is now time to relax and enjoy each other's company.

In many ways, the next stage of the method marks a loosening of the ties that have bound you in the first weeks. The highlight of this period will come when your dog takes its first walks off the lead and begins to explore the world on its own. Even before that eventful moment arrives, you can slacken the reins slightly here and there.

It is important that, at this stage, you feel happy and in control. Just as importantly, your dog must be allowed to progress at its own pace, and the worst possible thing you can do is to rush things. It would, for instance, be wrong to move on to the next stage with a rescue dog if it still cowers in corners or recoils at times of danger; the dog clearly does not yet believe it is safe.

My method, as you know, contains principles that must be incorporated permanently into your lifestyle. Most importantly, for as long as you and your dog are together, you must always reinforce your leadership when reuniting, as well as at times of danger and on the walk. Your behavior in these three fundamental situations will, slowly but surely, become as automatic as jumping into a car and driving down the road; you will become unaware that you are practicing it. When you reach this stage, you can think about eliminating the fourth pillar of my method, gesture eating, from your routine.

At this point I should stress that, before doing this, you must be satisfied that your dog has achieved the crucial breakthroughs of the first fortnight. The dog should have progressed to the point where:

- It behaves well at feeding time
- It responds well to all requests while walking on the lead
- It accepts its owner's leadership when visitors arrive at the home
- It has significantly reduced its repertoire when reuniting
- It is generally a relaxed and happy dog

Certain principles should, of course, remain in place when it comes to meal times. The relaxation of the gesture-eating rule is not a charter for bad behavior or bad eating habits. Under no circumstances should you now leave food on the floor at all times. And if your dog walks away from its meal, that meal must still be removed immediately. Dogs are always looking for chinks in their leader's armor, and you must remain vigilant in maintaining your control of meal time.

By now, you should know your dog well enough to read his behavior patterns. And if at any time you sense your dog slipping in its discipline, it is the easiest thing in the world to reintroduce the old routine. Gesture eating will always remain a powerful back-up tool for you when you wish to tighten up on your controls.

At the same time as you slacken the reins, you can also begin to be a little more relaxed about showing affection toward your dog. This, I know, always comes as a relief to those owners who feel somehow guilty for not fussing over their dogs during the early days of this method. It is important to state that I am not condoning a return to the situation where a dog comes bounding up uninvited onto an owner's lap. But, if your dog slides up alongside you or under your feet, as you relax watching TV or reading in the evening, there is nothing wrong with your now stroking it. Again, if the dog shows signs of disregarding your leadership, you can simply return to the old routine. Central to everything is a sense of flexibility. As long as you continue to be alert and open-minded about what is going on around you, you should be able to think your way through any situation that presents itself from now on.

Freedom: First Time Off-Lead

One of the purest pleasures of dog ownership comes when your dog is released to run free for the first time. The sight of a dog liberated to express its personality, athleticism, and natural exuberance never fails to bring a smile to my face. I have spent much of my life trying to help dogs lead lives more in tune with their natural instincts; here they are truly in their element. This is a key moment after all your hard work. If the first steps into your garden was the kindergarten, primary school was your first walk in the outside world, and secondary school was your first walk in the park on the lead. This is when you graduate.

As ever, you must exercise common sense and patience here. Your dog should always remain on the lead in built-up areas or near roads; many people fail to realize the intrinsic danger of letting a dog run free in such hazardous situations. Once in open space, however, your dog can be readied for release. The first time this is attempted, I recommend you go through a routine that once more underlines the principles established at home. The journey leading up to this moment has been a long and demanding one. It is this moment, however, that should pay back the dividends.

"Go Play"

Once more, preparation is important, so choose a place that is quiet, somewhere like a large local park. The key thing is to find a little isolation. When it is clear there are no major distractions such as other dogs or large groups of people, bring the walk to a halt as usual.

Step 1

Get your dog to wait, and remove its lead. Then walk with your dog at heel for a short distance, say twenty paces or so.

Step 2

Release your dog with a new instruction, such as "Go play." Be assured that this is the one instruction that will never need repeating; your dog will seize the opportunity with glee.

Step 3

Leave your dog to do its own thing for a while. Do not, however, let it stray too far, as this exercise is about building short-range control. When your dog is about ten yards away, ask him to come to you. Reward him with a tidbit of food when he comes.

Build up this work by increasing the distances to twenty, thirty, and finally forty yards. I do not recommend letting the dog more than forty yards away in any circumstance. Experience has taught me that this is the outer limit of the dog's comfort zone, and beyond this distance there is a loss of control. If anything unforeseen happens, you will be too isolated and may be out of your dog's hearing range. In case of incidents like this, where the dog may fail to hear your voice, I recommend you carry an ordinary whistle.

This is another moment to truly enjoy your relationship with your dog. Both you and your dog should be exhilarated by the sheer joy of running free. You can add to your dog's positive association with this moment even more by playing with it; there is no more powerful a moment. Show it that freedom is great, but that enjoying that freedom interacting with its leader is even greater.

Releasing the Pack

As with an individual dog, if you own a pack, you should ask them to wait after slipping them off their leashes. You should then ask them to walk closely together with you, before releasing them to "Go play." Again, it is important to apply your knowledge of the pack dynamic here. If there is a dominant figure within the pack, you should make sure this dog does not wander too far. The most effective way to do this is by working the dog hard through play and regular requests to come to you. By keeping this dominant dog

within a safe and controllable distance, you will ensure the rest of your pack remains in close proximity too. If you work your dominant dog hard at the beginning of the walk in particular, you can pre-empt having to watch it too closely later on.

It is very easy for an owner to find they are dictated to by their pack. Wherever I go for a walk with my dogs, I see owners who are afraid to relax, and who watch their dogs like hawks; the joy of the walk seems to have been lost. Of course, you must keep a watchful eye on your dogs, but it does not have to be such a strain. To my mind, it is far more relaxing all around if it is the dogs who are constantly keeping half an eye on your every move, rather than vice versa. And this is what I ask all owners to do.

The best way to establish this is to walk in a different direction from the dogs the moment they are released to play. You should remain calm, almost aloof, when you do this. When your dogs come, you should react as if this was inevitable, and reward the dogs with quiet praise. The worst thing you can do is to constantly shout instructions to your dogs as they play. Eventually, your stream of words will blur into one. Like the birds in the trees, or the sound of the cars in the distance, your voice will become little more than background noise.

My aim here is to establish the idea that it is the dogs' role to keep up with you rather than the other way around. This is not something that has to be done constantly. Again, it is vital that the walk is as natural a process as possible for your dogs, and they are able to explore their environment freely and safely. It is a question of striking a balance. If, for instance, your dogs suddenly begin charging off in one direction, you should start walking in the opposite direction, calling the dogs' names and asking them to come as you do so. In addition to sending out a powerful message, this process introduces an element of playfulness and fun as well, and the dogs will respond to this.

One owner with pack problems whom I have helped was Pauline, a lady with a very large pack of dogs and a particularly dominant Jack Russell, called Polly. Polly lived with six other Jack Russells and two German shepherd puppies in a beautiful country house near Hinckley in Leicestershire. It was a perfect home for a large pack of dogs, as Pauline's grounds extended over twenty-two acres—more than enough room for a decent run out each day. Pauline had read my first book, but had trouble imposing herself on her dogs during the walk. When I visited her at her home, I quickly saw that the problem was Polly, who insisted on leading the other dogs in her own direction.

After working with Polly through the key stages of my method inside the house, I took her and the rest of the pack out into the grounds. They were all off-lead, and scattered to the four winds instantly. I asked Pauline to stay at the back door with me, ignoring the dogs while they did this. It did not take long for them to realize that we hadn't joined them, and they soon drifted back to the house to find out why we hadn't come with them. At this point, I called Polly to me, extending the work I had done inside the house; she was agitated, but I waited until she relaxed. Once she had calmed down and had sat patiently for a short time, the rest of the pack quickly followed suit. I then asked Pauline to join me in leading the walk.

The important lesson was that we had showed Polly who was in charge of the hunt; she understood that she had to defer to me before we went anywhere as a pack. Pauline continued working along these lines after I left. Soon after, she contacted me to tell me she had a pack of beautifully behaved dogs. An already enjoyable daily walk had turned into something to treasure.

Close Encounters: Confrontations

It will not be long after those first, furtive steps into the world together that you and your dog will face up to your next big test: confrontations with other, less well-behaved dogs. Every owner faces this moment with dread, of course. We have all been there: another dog starts barking or, worse, charging in. The situation quickly deteriorates and becomes a vicious circle; as the innocent dog sees its owner become distressed, so it retaliates. Of course, if a dog has not yet been relieved of its leadership, it will do what it can to make that threat go away. The day these confrontations cease to be a problem will be the day the world is filled with considerate and compassionate owners. And I fear that day is about as close as the day we choose a dog for Prime Minister. So, you must be prepared to deal with this situation.

Just Walk Away

I do not believe in doing anything that is alien to a dog's nature. So, once more, my philosophy for dealing with this situation is drawn from the wolf pack. The simple fact of the matter is that, in the wild, wolves avoid confrontations with other wolves at all costs. The only time they will occur is if a lone wolf tries to infiltrate a pack, or an Alpha has been lost and a new wolf makes an attempt to join the pack. These, however, are exceptional situations. For the most part, wolves are intensely territorial creatures and avoid rivalries. My advice here, then, is that you do the same.

When confronted by an aggressive or overfamiliar dog, I suggest you simply move away, cross the road or take whatever action you need to get away from the situation. You should forget about apportioning blame or feeling aggrieved at the behavior of someone who is almost certainly less well enlightened; I do not wish to see dog dementia joining road rage as one of the sicknesses of the modern world! Just recall your dog if it is off the leash, return it to the lead, and walk away, using rewards to underline the fact it has made the right decision. Of course, this is a situation that will test your leadership skills to the limit. It is why I insist that owners spend such a lengthy period of time establishing their leadership credentials at home.

This area, I admit, is one that owners sometimes question. They ask: "What's wrong with two dogs playing with each other?" I am not for one moment suggesting that dogs should not socialize with other dogs. Far from it. As with humans, I think it's deeply unhealthy for a dog to live a solitary life away from its peers. The best comparison, again, is with a child. We all want our children to be able to play and interact with other children. But would we let them stop and play with every child they came across in the street? Would we let them go up to each child they meet and say: "Hello, let's play?" Of course we wouldn't. I encourage all owners to allow their dog to make a circle of friends. They can be dogs you regularly meet in the park, dogs belonging to your own friends, or through dogbased social gatherings. This, of course, also allows you to meet like-minded people yourself.

* * *

Of all the cases I have dealt with over the years, that of Amber and her owner Mavis best illustrates how seriously a fear of others can affect a dog and how hard owners sometimes have to work to communicate with their dogs. As it happens, it also exemplifies the potentially tragic nonsense some so-called "experts" present as advice.

Amber was an absolutely lovely looking dog, a beautiful little Sheltie, or miniature "Lassie." Ever since Mavis had taken her in, she had shown a tendency to become hyperactive and nervous in the presence of others. At home, she would jump around the furniture when visitors entered; but the problem was at its most acute when Mavis took her out for a walk. Amber simply didn't like meeting other dogs. She would try to run off and became extremely agitated if Mavis insisted on taking her toward another dog. Rather than improving as she was exposed to more and more dogs, Amber's problem was worsening. Mavis had resorted to taking her out for walks at night and at daybreak to avoid coming across other dogs.

Mavis cared deeply about Amber and did all she could to rid her of this fear. At one point, she grew so desperate she hired the services of a so-called "trainer" who worked with her in a park in London. He persuaded Mavis that Amber's problems could only be conquered by her confronting her fear, explaining that Amber simply had to face other dogs. I found it hard to believe this but, at one point, he actually dragged poor little Amber to meet other dogs. Of course, this only made matters worse. Amber just fell apart. She literally wet herself, and was reduced to a quivering, shaking wreck. Mavis eventually had to carry Amber home, she was so distressed, but not before her well-paid specialist had delivered his final analysis on her much-loved dog. He would not accept that he had failed and laid the blame elsewhere. "She's too nervous because she's been badly bred," was his dismissive verdict. If nothing else, I was determined to prove to Mavis that this man was a complete charlatan.

Mavis had become so embarrassed and concerned about Amber's behavior at home that she had stopped having friends to visit. When I turned up, I got an instant insight into why she had done this. Amber began bounding around the place like something out of a *Tom and Jerry* cartoon. It was clear to me that this was a particularly stressed dog and so I asked Mavis to come out of the room with me. Over a cup of tea in the kitchen, I explained the problem. Amber clearly believed she was in a position of complete power within the house. But as Mavis knew all too well, while Amber was Moby Dick at home, she was a minnow in the outside world. I set about the task of redressing the balance.

After a while, I sensed that Amber had stopped her leaping repertoire in the sitting room. I opened the door without speaking, but no sooner had I done so than she started all over again. Once more I asked Mavis to leave the room with me. A few minutes later, we returned to the sitting room and went through the same thing again. We went through this about eight times in all; Mavis's supplies of tea were severely depleted by the end of the day. Each time we left the room, the length of Amber's repertoire got shorter and shorter. Eventually, Amber showed the telltale signs of a dog that had begun to realize its situation had changed.

The key thing now was that Mavis build up on her leadership at home. My advice to her was that she did not walk Amber for ten days. She went along with this and used the time to build up her credentials as the Alpha in Amber's eyes. I'm glad to report that Mavis was rewarded with a radical change in Amber when their daily walk was resumed. Relieved of her leadership responsibilities, Amber was now able to walk down the road unthreatened. They are now able to walk about in open, wooded areas. Amber plays with dogs she regularly meets, and avoids those she doesn't without shaking, trembling, and falling apart. As for Amber being badly bred, Mavis has consigned the idea to history—along with the memory of the "expert" who suggested it.

The Pursuit of Prey: Chasing Other Animals

It is only natural that the walk is an area where many owners encounter problems with their dogs. The dog, after all, has its primal instincts heightened for the hunt, and its adrenaline levels are soaring. A dog's nature can easily run away with it at this point, often literally. As you progress to letting your dog off the leash, this is something you must be careful to guard against.

I have dealt with many owners whose dogs chase other animals while off the leash. I have come across sheep-worriers, horse-chasers, and rabbit-hunters. Each time I come across such a case, I begin by putting this behavior in its correct context. When a dog comes across another animal, and a herbivore in particular, its instincts are to set this animal in motion and then track its movement. Because of the nature of the prey it hunts, the wolf does not attack each and every animal it comes across. To begin with, it tends to hunt large animals and thus risks picking a fight with a foe that is more than capable of beating it. As other animals also tend to move in packs, the wolf's first priority is to get its prey on the move so that it can spot its weakest links. By observing the pack in motion, the wolf will be able to tell which are the slowest, and generally most vulnerable, members. It will base its attack plans on this information. If we translate this to the dog, we can see why they tend to chase other animals. On the rare occasion when a dog actually attacks, it is probably because it has spotted a weakness in that animal.

* * *

A typical example of this was a dog called Harley, whom I was asked to treat for my television program. Harley was a highly excitable Samoyed who lived with his owner, Andrea, in the countryside near Guildford in Surrey. Harley had several problems. In general, he took absolutely no notice of Andrea. The most severe—and potentially catastrophic—problem, however, was his habit of running wild on local farms.

Harley had begun chasing cows in particular. Andrea had

been visited by one local farmer who had warned her, quite rightly, that unless Harley was controlled, he would have to take drastic action. The last straw had come when Andrea had been out walking with Harley with another dog. Without any warning, Harley had suddenly frozen and then shot off like a rocket. Andrea had no idea what had caused this, but when she looked into the distance she could see someone schooling a horse at the end of a lunge line. The horse was moving around in tightly reined circles. Despite the fact that the horse was three fields away, Harley had shot off in pursuit of it. It had taken all the trainer's powers to keep Harley from attacking the animal.

It was soon afterward that I was called in. I recognized this scenario all too well; I began my first book with the story of my dog, Purdey, whose habit of chasing after animals had led to similar warnings from concerned farmers near my home many years ago. Just as I had been back then, Andrea was a responsible owner who was deeply concerned by this. She wanted to help Harley before it was too late.

I began by explaining Harley's behavior to Andrea. In particular, I was able to put the attack on the horse into its natural context. When Harley had looked across the fields to see this horse, he had seen what looked like a horse in trouble. Why else would it be moving around in circles? That must have been a sign of weakness, perhaps a broken leg. Whatever it was, it meant that horse was easy prey—and potentially, easy meat.

Andrea grasped this idea quickly. The key thing she now needed to do was to work hard at home with Harley over the coming weeks. She had to develop Harley's abilities at the end of a lead, and then she had to begin walking him past herds of animals in the countryside.

Andrea's goal was to become such a strong and convincing leader that Harley would not even contemplate the need to go on the hunt. She had to make him so confident in her abilities that he trusted her to provide food. I warned Andrea that she had a long and difficult task ahead of her, but Harley responded well on that first afternoon. The future will be down to Andrea's determination to succeed. I'm sure she's more than up to the challenge.

Traveling Further Afield: Trains, Planes, and Automobiles

I often compare learning my method with learning to drive a car. Owners progress from basics such as learning the controls to dealing with life on the open road, coping with encounters with other motorists, and so on. I often warn owners of the dangers of heading out on to the motorway when they have not learned yet to navigate their own driveway. When it comes to taking your dog out in your car, or traveling further afield by train or plane, this advice applies in the most literal sense.

Once again, it is easy to see why this area is rife with problems. When a dog jumps with its extended pack into the back of a car, or on board a plane or train, it enters an environment that is, to all intents and purposes, a miniature and mobile version of the den. When this portable den starts to move off in a way it cannot control—or even comprehend—the dog's anxieties multiply with each passing moment. Out on the road, it is assailed by sights and sounds it cannot understand or reach, and is convinced they are going to harm the pack members for which it is responsible. Is it any wonder a dog barks, leaps around, and generally flips its lid when it gets into a car? For this reason, it is more imperative than ever that things are not rushed. I do not recommend that you take your dog out into the car until well into the second or even third week. Such is the level of trust required that to do so any earlier risks your losing much of the ground you gained in the earlier stages.

* * *

I have dealt with a host of owners who have had particular problems with their dogs causing what I call "car chaos." The case of Ziggy illustrates where owners can typically go wrong. Ziggy's owner, Carol, was a highly successful construction industry professional from Nottingham. Ziggy was a former rescue dog, and, like so many of these dogs, he was edgy and nervous and could get extremely lively at the best of times. He used to charge at people. He was also extremely attached to Carol and would follow her everywhere. She liked to take Ziggy out and about on sites with her, but had serious trouble keeping him calm in the back of the car. Whenever Carol went anywhere near her car, Ziggy bolted toward it first. Once he was inside, he would jump around almost continuously. He would calm down when they were on motorways. But in built-up areas, when they were surrounded by other cars and pedestrians, Ziggy would bark incessantly at the outside world.

Carol read my first book and began applying my method in an attempt to rid Ziggy of all his behavioral problems. Her greatest ambition, however, was that he would be able to travel around with her. Rescue dogs, as we know, need to be treated at their own pace. They are, after all, the equivalent of children with learning disabilities, and allowances must be made. Carol had been delighted with Ziggy's response in general. He had learned to respond to her "Thank you" at times of perceived danger, and, in the space of three weeks, he had become a much calmer dog. In the car, however, his problems remained much the same as before. It was at this stage that Carol called me in.

It did not take me long to work out where Carol had taken a wrong turn. She admitted that, while she had held back from taking Ziggy out for a walk until the second week, she had been taking him in the car from the outset. It is easy to see why Carol had done this. For a start, she did not want to be separated from the dog that she loved. And, somewhere, I'm sure she felt that her car was simply an extension of her home.

Unfortunately, her honorable motives had been totally lost on Ziggy. For all the progress Ziggy had made in the home, he had not fully accepted Carol's leadership. In addition to this, Carol was getting agitated and upset with Ziggy's behavior in the car, so to his mind she could not successfully lead the pack within the car either. All this had become a block to his accepting her as his superior. It had also made him associate his owner's anxiety with the car—a dog is never going to think it is the cause of the problem! As a result, Ziggy was as hyperactive a passenger as ever.

When I visited Carol, I worked with Ziggy on my own to begin with. I got him to wait alongside the car, before he jumped in. It was important he begin to understand that he did not decide when he entered the car. When we went for a short drive, I simply sat beside him quietly. My message to him was that I understood and could deal with all the dangers on the other side of the car's windows. He did not need to worry; there was no requirement for him to do anything other than enjoy the scenery.

Ziggy responded well, calming down significantly, but it was up to Carol to take control of this now. She had to build up her control of their drives together, so I got her to work on different elements of it. Carol began by getting Ziggy to stay at the door of the car. He had to learn to wait for her instruction to get in. From there, he was harnessed in the back of the car. Carol then took short journeys, building up her control of him. At first, these trips were literally the length of the drive and Carol would ease the car along, silently reassuring Ziggy all the time that he was safe in her hands. When he began to calm down, she slowly lengthened the journey. Soon, Ziggy was traveling around with Carol everywhere again—only this time he was able to take in the views.

PART SIX:

Days 22–30

The Counter-Revolutionaries: Leadership Challenges

By now, you should have made huge strides. You will probably be feeling pretty pleased with yourself—and rightly so. The biggest mistake you can make, however, is to imagine that it is all plain sailing from here on. Choppy, capricious, and potentially disastrous waters lie ahead, and it's important to keep a firm grip on the helm.

By far the most common calls I get at this stage come from owners who suddenly find their dogs displaying strange and seemingly uncharacteristic behavior. Typically, they call me with cries of "My dog has started eating the contents of the washing machine," or, "My dog is chasing his tail." The phone call invariably ends with the phrase, "He has never done this before. What on earth is this is all about?" The answer is a relatively straightforward one. Their dog is, once more, testing its owner's leadership. And, contrary to the impression the owner may have had of themselves, their dog may have decided that they are not measuring up.

At the heart of the matter are three factors that come together at the same time. Firstly, you must remember that your dog is breaking the habit of a lifetime. Until the introduction of the new regime a few weeks back, they were convinced of their status as leader. The residual memory of this lingers, particularly in homes where the dog may have ruled the roost for years. Secondly, the first weeks are, in many ways, a honeymoon period. Much like someone settling into a new school or working environment, the dog's focus and energies have been tied up in dealing with an often bewildering set of practicalities. By now it has found its feet and, as a result, is beginning to revert to its natural character once more.

The final, and most important, factor is that you will now have relaxed. Although this is understandable as the first few weeks are hard, demanding work, this is no time to rest on your laurels. Because now is the time when your dog is most likely to spot a chink in your armor.

Crime and Banishment

As you know, my method entails a quiet revolution. But what if you are dealing with a dog that is still determined to cling to power? The answer is that, sometimes, overthrowing a dictator requires more visible and stronger tactics. Again, the strongest tactic possible is drawn from the example of the wolf pack. Faced with unacceptable behavior or insubordination in the wild, an Alpha wolf will physically eject a transgressor from the pack via aggression. The ejected wolf will be held on the outskirts of the pack until it has either learned its lesson or has left. Any attempt to reclaim its place prematurely will be met with more aggression.

At home, an owner cannot physically do this, nor should you want to. What you can do, however, is be rigid of intent and use the same principle. The more severe the challenge to your authority, the more severe your signal to the dog must be. And there is no more severe a signal than banishment from the pack. As with all the most powerful weapons, this must be used sparingly. It serves two purposes: when it is dispensed, pack members sit up and take notice, while the knowledge that this may happen also acts as a deterrent.

Banishment is something owners find difficult to carry out, I know; but it is vital that they remember to work with the dog's basic instincts. On the one hand, this method respects the dog's need to challenge its owner's leadership. Yet, at the same time, it uses the knowledge that a dog needs to be part of a pack environment in order to survive. These twin forces will combine to produce the right outcome.

Frequently Asked Questions

When Do I Use Banishment?

As ever, there has to be a huge flexibility here, but I recommend it is used whenever your dog is openly defiant. If, for instance, you invite the dog to come to you and it merely looks at you and ignores the request, your dog should be cut off from you for the rest of the day. It should still be fed using gesture eating, but it should not be walked or played with, and should not even have its presence acknowledged for those twenty-four hours.

In my experience, this is often necessary with remedial dogs. They can look at you with contempt, as if to say, "Who do you think you are calling?" You have to make that dog question its decision. And the way to do this is to ignore it for a day. This may sound severe, but a way is needed to transmit your message, and what are the alternatives? The old-fashioned method would have been to shout and holler and shake the dog. How much of that is going to mean anything to it? Isn't it far better to use a language it understands and threaten it with a sanction that really impinges on its quality of life?

What if My Dog Really Misbehaves?

In really extreme cases, I ask owners to ignore their dogs for two, three, or even four days. This happened to me once with my own much-loved dog, Sasha. At the time my springer, Molly, was coming into season, so it was a time of heightened tension within any pack. One day, as Molly walked past, Sasha grabbed at her in a violent way and a brief confrontation ensued.

Such times are testing for any leader. It is imperative that a show of strength is made—and that it is one that the dog understands. I was not going to have behavior of this kind in my pack so I drilled Sasha back under the table, using my eyes alone. I didn't speak a word, but she knew I wasn't happy with her. I then decided I was going to keep her there for two hours. So I stayed in the kitchen, clearing out cupboards. Every time Sasha attempted to move out from under that table, I glared at her and she went back. She knew she was in trouble, and knew that her leader was in charge. After two hours I decided to let her out, but again I did this without looking at her or speaking to her. When she made a move to go out, I let her go. She crept out, then came to my side, trying to apologize. I ignored her and continued to ignore her for four days. I showed her this was totally unacceptable.

During this time, I took Sasha out for walks with the others, but I kept her out of the group again using my eyes. Whenever she tried to rejoin the pack, I kept her out with a glare. This was no easy thing to do; I found it as hard as any owner would have. I loved my dogs and I wanted them to be happy together as a pack. But at the same time I did not want my dogs to fight, and this was the only way of ensuring they didn't. Ultimately, the ends justified the means. After four days, I welcomed Sasha back into the pack and there was no more trouble between her and Molly. The message had been a hard one, but it had got through.

* * *

Challenges to leadership can manifest themselves in all manner of ways, as two examples from my casebook may help illustrate. The first involved a family with three dogs: Stan, Ollie, and Buster. This trio lived with their owner, Debbie, in Peterborough. They had made great progress in applying my method; the dogs behaved very well at home. Gesture eating had been successful; so was reuniting after a separation. The dogs' repertoire had reduced to such an extent that the five-minute rule had evolved into the one-minute rule. The only problem lay on the walk, when the dogs continued to pull on the lead and charged into a boundary fence as strangers passed by. Debbie, like so many owners I meet, was determined to iron out this last, frustrating problem. She needed a hand with the final phase, and so requested my help.

Time and again, I arrive in these situations to discover that the problem area has arisen outside of the home. This is understandable; we all feel far more in control within the confines of our own home. Outside, in a world full of unpredictability, our confidence levels are diminished. Unfortunately, this is something a dog seizes on instinctively. It was clear to me that all three dogs were not yet fully convinced of Debbie's leadership credentials. But it was Buster who demonstrated his doubts in the most memorable—and amusing—way.

Debbie had made one classic mistake. Because Buster was smaller than the other two dogs, she had imagined his problems were smaller. Buster is what I jokingly call a "multi-pedigreed" dog. As a result, Debbie had concentrated on her Labradors more than on him. She had thought Buster would require less effort; how wrong she had been!

Until now, the dogs had shown every other sign of believing in Debbie's leadership. It was when I appeared on the scene that Buster really showed the true nature of his feelings, and staged his last stand. Debbie and I sat in a conservatory, where Stan and Ollie, the two Labradors, were fine. Buster, however, was trotting around the place, looking very preoccupied with things. As I sat and listened to Debbie's account of the situation, studiously ignoring all his posturing, Buster started barking at me. Debbie immediately came out with the line: "He doesn't normally do that." I got her to lead him outside the conservatory to the garden. Out of the corner of our eyes, we watched Buster do a circuit of the garden. He came up to the glass, barked, then calmed down again.

When we let Buster in, he scampered over to a favorite wicker chair and was calm there for a short while. Then, suddenly, he started pawing rhythmically away at a cushion. Again, Debbie was bemused: "He's never done that before either." His barking had failed so, clearly, Buster was going to try something else.

Once more, Buster was exiled from the conservatory. And once more, he went off to do a circuit of the garden, clearly planning his next move. We soon found out what that was. The next time he was let back in, he darted behind some of the large plants Debbie had kept there. For a few moments, all was peace and quiet. Then, as if in slow motion, Debbie and I saw one of the plant pots moving ever so slowly and slightly. When we looked a little closer, we saw that Buster was there chewing furiously at one of Debbie's precious plants. I must admit we had both found Buster's previous attempts to assert himself funny. This time we couldn't hold back and both let out a giggle.

Buster was telling Debbie that, for all the progress she had made, he still didn't accept her as leader. Dogs are sensitive, and Debbie confirmed to me that she didn't really see herself as an authoritative leader. This was the key: her belief level was not high enough, therefore the dog's belief level was not high enough either.

Eventually, Debbie prevailed through showing strong, silent leadership, and Buster gave up. Within a few days, by going back to basics with the principles of Amichien Bonding, he, like Stan and Ollie, had accepted the new order. The conservatory had been Buster's Little Bighorn, the site of his brave, but futile, last stand.

* * *

A dog can show it is unconvinced about leadership in a variety of ways. Another owner I visited was completely baffled by her dog Danny's behavior. The owner, Sarah, was three-and-a-half weeks into applying my method on Danny, her terrier. Until then, things had gone along a predictable, successful course. In the home, in particular, Danny had responded encouragingly. His only weakness came when people passed outside his front window. He would leap onto a chair and then onto the windowsill. Sarah had no ornaments left there, so animated were Danny's protests at passers-by. Even here, however, his behavior was improving. His protests were becoming less dramatic and shorter in length.

Then one day, as Sarah played with Danny and a toy in the garden, he did something he had never done before. When she threw him the ball with a bell inside it, Danny took it and ran into a corner where he proceeded to shred it. There was a manic energy about him; it was, Sarah said, as if he was very, very angry with her. She rang me for help.

I recognized Danny's case as a classic example of a dog coming back at its owner and mounting a serious challenge to her leadership. I explained to Sarah that, effectively, Danny was saying: "You have applied for the job of leader, and I have agreed that you can have it. But now you are not measuring up to the challenge. And that is really annoying me."

Sarah clearly had to regain the upper hand here as quickly as possible. I got her to remove all the toys from the garden and stop playing with Danny for a few days. He was very angry with her because, when Sarah had resumed normal play after a few days, he had once more shredded a plastic toy. While this was going on, I got Sarah to reintroduce gesture eating and to become more formal in her dealings with Danny once more. Within three weeks he had become calm once more. The counter-revolution had been quelled, and Sarah was determined that Danny would have a leader he could trust in the future.

* * *

A dog has no understanding of human status; it is of no relevance to it that its owner is a celebrity or a supermarket shelf-stacker, a president or a postman. A dog judges its leader by his ability to lead convincingly, and you cannot con a dog into believing this if it is not true. If there is a weakness, the dog will latch onto it. As a footnote to this section, it is worth mentioning the example of a dog that had its own small place in the history of the twentieth century. As a young girl, I can remember seeing footage of Adolf Hitler with his pet German shepherd. The two were playing at his Bavarian retreat. At the time, of course, I had no real understanding of Hitler or his significance in the world; my main interest was dogs. What struck me about what I saw was how unhappy his dog appeared.

On the several occasions I have seen the footage since, I have felt exactly the same thing. I now know the monstrous truth about Hitler. And I know how, through fear and intimidation, he prevented anyone from challenging his leadership. Few humans found themselves able to stand up to his overwhelming, bullying personality. I'm sure he treated his shepherd with the same arrogant imperious manner. Yet the dog was having none of it. Hitler hadn't even begun to convince him he was a real leader. I wonder what inner weaknesses the dog sensed in him?

All Change: Coping with Upheaval

There will, of course, be times when your family's shape and surroundings change. Children may grow up and leave home, family members may pass away, partners may separate. This may easily unsettle and upset a dog. Equally, your family may face the upheaval of moving. The prospect of adapting to new surroundings will be as challenging to the family pet as it is to everyone else.

It is easy to see why such moments can throw a dog off kilter. Dogs equate change with a challenge to the existing order and need to see order restored as quickly and emphatically as possible. So, in each of these circumstances, the key is to remain calm yourself, and to remain the inspiring, reliable leader your dog has grown to believe in. If the change of scenery or personnel causes major behavioral problems in your dog, revert to the early days of the Amichien Bonding process, reintroducing gesture eating and cutting out walks until your dog has once more accepted its status within the home. As with so many aspects of applying my method, it is a case of using patience, understanding, and plain common sense.

Long Separations

At the opposite end of the scale from the normal day-to-day separations I have dealt with in the earlier section of this book are the lengthy separations that come when you take long holidays or travel on extended business trips. Anxious owners naturally worry about leaving their pets for long periods of time, fearing that the separation may affect their relationship somehow. In my experience, there is little to worry about. In fact, as I have already explained, a dog has no sense of time so will not differentiate in any significant way between an absence of a few hours or a few weeks. In both instances, the important thing is that the dog re-establishes itself in the home and that you reassert your leadership immediately on your return.

New Arrivals

Nothing creates a more joyous transformation within the home than the arrival of a new baby. New parents often worry about the effect the newborn family member will have on their pets. The fact is there is no reason why it should have any real impact at all, other than a happy one. A dog that has been relieved of the pressure of leadership will take its lead from you. It will, naturally, have its curiosity aroused by the demanding—and occasionally deafening—new arrival. You must deal with this in the way you now deal with all unexpected noises: by thanking the dog for its interest and taking charge of the situation yourself. As ever, the key is not to make a huge commotion.

* * *

As a cautionary tale, I will mention one family who, unintentionally, made matters worse rather than better. Craig had spent six years living with his dog Joe before he met and had a child with his girlfriend. When the baby arrived home for the first time, the couple did what most sensible parents do: they made a big fuss of Joe, the older "child" in the home. Their intentions were, of course, entirely honorable; they did not want him to feel left out or overshadowed by his new "sibling." The problem was that, in this case, they were doing the exact opposite of what they should. Joe already believed he was the leader of the house. By their behavior, the family was vesting even greater power in him.

The baby was six months old when I was called in to see Joe. Joe's stress levels by then were so high, I was seriously worried that he was going to have a heart attack. Joe put up a huge battle before relinquishing his leadership. His repertoire went on for three-and-a-half hours, the longest I have ever known. I explained to the couple why Joe was unable to cope with these altered family circumstances. Craig and his family understood and began to implement my method. When I last spoke to them, they were still struggling but their resolve was strong. I'm sure they will get there in the end.

PART SEVEN:

Day 31 Onward...

Keep an Open Mind

If I have learned anything in the years that I have been working with dogs, it is to expect the unexpected. It is, if I am honest, the most enjoyable element of my work. From the tail-chasers and paw-chewers to the dogs with phobias about phones and fireworks, I have relished meeting each challenge. No two days can ever be the same—nor would I want them to be. There is, however, a common quality that is shared by all the owners I meet: all of them care deeply for their dogs. And all of them are prepared to go that extra mile to give those dogs a better life.

The method I have outlined in this book will provide every owner with the foundations upon which to communicate with their dog. But real, lasting success will come to those who add some of the other, almost instinctive, qualities that I have seen demonstrated over the years. They are not special gifts in any sense—anyone can have them. You should aspire to include these qualities too. If you do, I feel sure you will enjoy an even richer and more rewarding life with your dog.

I like to believe that my ability to keep an open mind has brought me a long way in my dealings with dogs. If I had been closed to new and original ways of thinking—no matter how extreme or far-fetched at times—I would not have made the breakthroughs that underpin my work. It is an attitude that I believe can serve all dog owners.

A case I dealt with a year or so ago is an interesting case in point. I was asked to visit a family of five who owned a lively dog. The family had been applying my method and had made tremendous progress. The problem centered on the family's teenage daughter. The dog had been with them since it was eight months old. It was extremely well behaved—so much so that, when I arrived, I wondered whether there was very much for me to do. But then the daughter arrived and everything changed. The dog backed away and ran into the kitchen, where he cowered in a corner, trembling.

The mind works overtime in situations like this. "What on earth had this girl done to the dog to produce this sort of reaction?" I asked myself. In other circumstances there may have been cause for concern. I might have wondered whether there had been some abuse going on here. But there was no question of the girl having been unkind to the dog; she simply thought the dog hated her.

It was as I spoke to the girl that I was struck by the perfume she was wearing. It was not overly strong but, equally, it was hard to miss at close range. To a dog, such smells are deeply confusing. It knows the aromas of its pack members and associates strongly with these. Could it be that, since the daughter had started wearing perfume, like any average teenager, the dog's attitude toward her had changed? I had to tread carefully here, so I asked the girl whether she always wore the perfume. She was slightly offended by the question but eventually understood why I had enquired. I asked her to stop wearing it around the house for a week, and also asked her to stop using perfumed soap. Within two weeks the dog was reacting to her in exactly the same way as he reacted to the rest of the family. She did not smell any less delightful when she went out at night.

Be Adaptable

Most of the dog owners I meet are committed, caring people who put the welfare of their animals at the top of their priorities. They will do anything to ensure their dog lives a happy and fulfilled life. Unfortunately, however, the world is not a straightforward place, and there are obstacles that sometimes cannot be overcome. It is important that, for this reason, you are able to adapt to the surroundings around you. And sometimes that involves making allowances for those people who do not share our view of the world.

* * *

When I think about this aspect of dog ownership, I often recollect the determination of Linda, a nurse from Yorkshire. Linda lived with her two Cairn terriers, Milly and Molly. Unfortunately, her next-door neighbor was the owner of a very badly behaved and aggressive dog, and could not have been further removed from the kind of caring dog owner that Linda represented. The neighbor's dog would bark at Milly and Molly, and would fling itself in a rage against the fence that separated the two gardens. Milly and Molly would react to this aggression, and the noise the three dogs made would be deafening.

We can no more protect ourselves from such undesirable neighbors than we can from the onset of the seasons every year; it is a part of life. It was clear this was not an environment that was going to allow Linda a completely free run with her dogs. By adopting an adaptable approach, however, she performed minor miracles. She began by erecting a formidable, six-foot fence between her and the garden next door. She then set about introducing my method to her dogs, concentrating her work inside the home. When it came to the time to take her dogs into the garden, she combined the work she had done on perceived danger and the come and recall to good effect. When the dog next door began his barking, she called Milly and Molly to her, rewarded them with tidbits, and thanked them for alerting her to the danger.

By teaching Milly and Molly to believe in her leadership absolutely, she soon persuaded them to ignore the dog next door completely; gradually his charges and protests became less intense. This dog was very unhappy. The more Milly and Molly had reacted to him, the more animated he became. Denied any interaction, he eventually gave up. I like to think we helped him indirectly too.

Fortunately, this story has an even happier ending. A few months after I met her, Linda got another job in another part of Yorkshire. She was delighted to find a house only three doors away from her brother. It was hardly surprising that the dogs' behavior changed almost immediately. Linda described them as transformed, joyful creatures. In stark contrast to her previous neighbor, the people in the two houses separating her from her brother happily allowed them to build a channel which allowed the dogs to run from one house to the other. I like to think someone, somewhere was rewarding Linda for her perseverance. . . .

Think Holistically: See the Big Picture

The most common misconception I come across among owners of multiple dogs is the idea that the problems of one dog can be treated in isolation. This is simply impossible: the pack must be treated as a whole, for a variety of reasons—not least the fact that far too often the owner is guilty of mistaken identity. I often come across cases where the root cause of the pack's problem does not lie with the dog they imagine, but elsewhere.

* * *

So it proved in one particular case I was called in to deal with, that of the dogs I refer to as Puppy X and his uncle. This was another of the MI5-style operations I have been involved in. Once more, I had been sworn to secrecy on the case by the client. And once more the client came from the world of breeding and show dogs.

The family concerned were the owners of a large, and very successful, show kennel. They asked me to tackle a problem with one of the thirteen dogs they had at the time, an eighteen-month-old dog. The family had a long and well-regarded record at dog shows. As a youngster of six or nine months, the puppy had done well in the show arena. Now, at the age of eighteen months, however, his demeanor was totally different. The dog would back off from judges whenever they approached him. Sometimes he would sink down on his haunches. The owners told me their dog had become really unhappy, and as genuine dog lovers they could not countenance this.

This behavior in dogs is far from uncommon within the show world, particularly if the owners grow anxious and frustrated within the ring. This nervousness can transmit itself to the dog—with predictable consequences. This was patently not the case with these owners, as they quite rightly regarded their dog's welfare as far more important than a winner's rosette. They told me when I got there: "It won't bother us if he doesn't go to another dog show in his life, we just want him to be happy."

I arrived at their property and was led into a room where three dogs were waiting. It was fairly obvious which one I had been called to deal with. Puppy X's demeanor was very defensive and he was edging toward a comfortable-looking cage in the corner. Clearly this was his own space, his personal den. But it was not long either before I had spotted that the cause of the puppy's anxiety was also in the room. And that was his four-year-old uncle.

I had gone into the house in my usual manner, refusing to engage with or acknowledge the dogs there in any way. While the youngest dog cowered and the second dog, clearly the mother of the pack, just stared at me, the uncle could not cope with my presence and went into an extraordinary repertoire of barking and jumping. The more he protested, the more I made a point of ignoring him. It was as if he was saying: "Who does she think she is?"

What was really revealing was the way in which the uncle treated his owners. As his repertoire continued, he jumped up onto the knee of the father of the household. I asked the owner to repel the dog, but he found it difficult because the dog's anxiety levels were so high. It was clear to me then why Puppy X was not performing well within the show ring. Quite simply, the dog had little or no regard for its owners because of their status within the home. There was only one leader there—and it was not the owners, it was the uncle. I had no doubt that the jumping up I had seen was a regular event. And that, until now, the uncle had got exactly what he wanted from his owners.

I explained my analysis of the situation to the family. And I told them that, rather than tackling the younger dog, my job would begin with the uncle. I worked with the three dogs that were in the room to begin with. Eventually, the uncle's repertoire came to an end; it must have taken about half an hour of over-the-top behavior before he gave up. He was obviously a highly intelligent dog, but he was soon coming to me.

While the uncle had been going through this process, Puppy X had remained in his own space. It was clear he was wary of getting involved. But it was also obvious that he was waiting to see how events panned out. When, eventually, the uncle calmed down and began responding to my requests to come, Puppy X slowly inched his way forward. Seeing his uncle responding gave the younger dog encouragement. And for the first time he, too, began to come toward me. He had sensed a change in the household's hierarchy and was reacting to it.

My next task was to introduce the rest of the pack to the method. I began by asking the owners to bring in the two puppies. The moment they reappeared, the uncle went into his repertoire once more while, at the same time, the younger dog retreated to his cage. I waited for the uncle to calm down once more before getting the puppies to come to me. This was—as is often the case with packs—a long and laborious process. Owners often feel they are taking one step forward and one step back, but in fact they are always moving on.

I could not work my way through the entire pack in one sitting. Besides, the owners told me there were clear groupings within the pack, so I emphasized the need to work with these mini packs separately and left them to it. The owners grasped the principles immediately. They worked hard over the coming weeks, and were rewarded with huge improvements in their pack's behavior. Puppy X lost his haunted look within three days. Within the canine hierarchy, the uncle remained top dog but, when it came to his human owners, he had relinquished responsibility. And that was a relief to everyone.

Perseverance Pays

We are all assailed by self-doubt now and again. If ever I question the worth of the work I am doing, I think of some of the hundreds of success stories with which I have now been associated, either directly or indirectly. One of the most heart-warming of these provides a perfect illustration of the importance of perseverance, and not giving up on a dog.

The story was told to me by a Customs Squad dog handler, who approached me and introduced himself at Cruft's. He explained that he had been using my method in his work with a rescued spaniel being trained as a sniffer dog. Customs call these dogs "second chancers" because they have—somehow—lost their first chance and ended up in the rescue center. The Customs and Excise Squad offer them a second opportunity. For many dogs, it might be their last.

It looked like this spaniel had also blown his second chance. He had been earmarked for work at Dover docks, where he would search for drugs and contraband, but he had quickly proven too hyperactive to concentrate on the task at hand. When let loose in a cargo container or the back of a haulage truck, the dog would skitter around in a haphazard fashion, sniffing in one corner at one moment, then the other corner the next. There was no method to his madness at all, and this was clearly not going to be accepted within the professional environment of the Customs and Excise Squad.

Again, this behavior was no real surprise given the dog's background. Fortunately for him, he had found a handler who was willing to go that extra mile. The handler told me that he had applied the method from my first book. The primary example of a rescue dog there was of my Jack Russell, Barmie. The slow, patient approach I was required to take in restoring Barmie's confidence in humans had provided a blueprint for the handler during his first few weeks with this dog. His patience and perseverance had soon paid off. Within a few weeks, he was working with a completely different dog. Rather than leaping around all over the place, the spaniel had become much more methodical. He would work his way through the boxes, sniffing them out thoroughly before moving on to the next one; there was nothing random about his work anymore.

The handler's hard work quickly paid spectacular dividends. Soon after he had been put to work at Dover, the dog boarded a large truck recently arrived in England from the Continent. The old dog would have bounded around haphazardly. Now it worked its way through the cargo, until its attention was drawn by a particular package. As he became more animated his handler moved in. The dog had hit the jackpot: the truck contained heroin worth half a million pounds. It was one of the biggest drug busts of the year. He had earned his second chance—in some style.

* * *

In the course of my work, I am asked to deal with a variety of dogs—from the aggressive and overexcitable to the sensitive and nervous. When it comes to helping them and their owners overcome their difficulties, I treat all of them as equals. Occasionally, however, I come across cases that push me, my method, and the owner concerned to the outer limits. To borrow a phrase from George Orwell, some dogs are—in terms of a challenge—more equal than others.

None of the diverse collection of dogs I have encountered was more equal than the pet that became known as "Damien, the Devil Dog." He represented perhaps the toughest test any owner has faced using my method. And the way in which that owner overcame the challenge should offer solace to anyone who finds themselves struggling to implement my ideas.

To protect the privacy of its owner, I will refer to this dog, a Welsh springer, as Tim. Tim displayed many of the common problems found in dogs; what distinguished him was that he was extreme in all of them. He pulled on the lead for all he was worth; he reacted to the arrival of the morning post by ripping it up; his aggressive streak was so bad that he had to be muzzled when visiting the vet. Tim's behavior had driven a wedge between Jenny, his owner, and her family. She was married with grown-up daughters, and the rest of the family were agreed that the dog had to go. Jenny had tried all she could to improve his behavior, but nothing had worked.

Every now and again, I come across a dog that has the look of a natural Alpha—the sort of animal that, cast into the wild environment, would rise without any challenge to the top of its pack. Tim was one of these dogs. He was not that big—he didn't weigh more than fifty pounds—but he had immense presence. Jenny, in comparison, was a very gentle lady. It was easy to see why the hierarchy had established itself in the way it had.

Despite the daunting task ahead, Jenny was absolutely determined to help Tim. I was very touched when I first spoke to her. "How can I let him go—where would I put him?" she asked me plaintively. The dice were loaded against her in many ways. In asking people to adopt my method, I am asking them to take a leap of faith. They must believe—no matter how improbable it may seem—that by applying my principles, their dog's behavior will improve. Unfortunately, Jenny's family were unwilling to share her faith. Her husband wanted Tim's behavior to improve too, but he simply could not see that adopting my method was going to help. In effect, Jenny was on her own.

As I have stressed throughout this book, there can be no precise timetable of events to my method. A lot of people want to wave a magic wand and have the problems disappear overnight. Life does not work that way. Yes, some dogs relinquish their leadership with barely a protest. But others fight tooth and nail to cling on to a status they believe to be rightfully theirs. No dog ever put up a fight to compare with Tim.

Jenny began applying the four elements of Amichien Bonding in the normal way. It was clear from the beginning that Tim was not going to give in. Any ideas Jenny had of heading out for a walk after one week were quickly forgotten. It was only after two weeks of working with Tim that she saw her first sign of progress. He had begun calming down during his repertoire. A week after that, Tim came to Jenny when she asked him to—something that had never happened in the house before. Until then, Tim had looked at the humans in the house as if they were total idiots.

Progress was, however, painfully slow. Jenny often felt that, for every step forward, she was taking two back. Soon after she had begun to get Tim to come to her, he began rebelling. He would run upstairs and pull all her shoes out of the bedroom wardrobe. She had to put a gate there to stop him trashing the room. To be fair to Jenny, the situation was not helped by the fact that the rest of the family were not applying the method. She was also a lady with a lot of self-doubt and, quite simply, Tim had not even begun to be convinced by her leadership. The pressure this was putting on Jenny was considerable. And at times, inevitably, she cracked. She would phone me every now and then in a terribly distressed state; she would cry and sob over the phone. Sometimes we would talk about Tim for an hour.

This struggle went on for two months before it entered its next phase. By this point, Jenny hadn't even progressed to taking Tim out for a walk, and he had shown signs of tightening rather than loosening his grip on the leadership of the pack. One day when Jenny invited Tim to come, he simply turned his back on her. She rang me almost immediately, once more distraught at his defiance. My message to her at this point was simple: this is a battle of wills, and your will must be stronger. "If the dog ignores you, then you must totally ignore it—for a whole day."

So, the next day, Jenny ignored Tim completely. Her only contact with him was at meal times, when she gesture ate in front of him. When, the following day, she once more tried to get Tim to come and he turned his back on her again, she responded by ignoring him for the next two days. This process went on until Jenny ignored him for six whole days. There was no aggression or anger in her body language. She simply got on with her day-to-day life without acknowledging the dog in any way, shape, or form.

The reaction of Jenny's family and friends to this was predictable. They thought she was crazy. Some told her she was being cruel to the dog; others suggested she simply give the dog "a good hiding"; choke chains and even an electric collar were suggested. Jenny, to her credit, refused to quit, however. And she was rewarded almost immediately after that six-day standoff came to an end.

The next day, Jenny called Tim to her and, rather than turning his back again, he moved—very slowly and gingerly—toward her. It didn't seem like much of a breakthrough, but, in the context of the war, it was a significant battle won. Until now, Tim had always had Jenny dancing to his tune. Now she had the upper hand—and it was vital that she did not lose it.

About an hour after this had happened, Jenny asked him to come again and he refused. She called me immediately. "What's happening here?" she asked. "He's still testing you," I told her. "Be firm." Jenny's response was to return to ignoring Tim completely, this time for four days. It was not as severe a consequence as the previous time because Tim had made a small move in the right direction. If he was a smart dog—and there was no question in either of our minds that he was—Tim would understand this and react accordingly.

Sure enough, when Jenny called Tim at the end of the four days, Tim came to her readily. Now I asked Jenny to really capitalize on her gains. Rather than rewarding Tim with some attention, I asked her to ignore him for a day. Jenny, of course, thought this was harsh, but I explained that she had to ram home the message that she was the leader. And now was the time to do it.

The morning after she had ignored him for the whole day,

Tim came readily to Jenny the moment she called him. I asked her to leave him then and call him again in the afternoon. Again he came. Over the next three days, I asked her to gradually build on this, calling Tim one more time each day. By the end of the fourth day she had called him four times—and he had come every time. She was even gently stroking his head as a means of praise.

Four months after she had begun the process, the turnaround had finally begun. Tim had at last begun to accept that Jenny was, after all, a formidable leader—someone worthy of respect, and someone he would be prepared to follow. I must admit it felt very odd when Jenny stopped calling me shortly afterward. I'd grown so used to hearing her voice at the end of the telephone. Almost every day I was tempted to call her, but I knew I had to leave her to build on this success. It was five more weeks before I heard from her again. She couldn't mask her excitement as she delivered her opening words. "The devil dog is no more!"

At first, even I was skeptical. I was waiting for Jenny to tell me that he'd been behaving well for a day. It turned out that, for the past week, Tim had transformed into a lovely dog. He had been out on a walk each day and had behaved impeccably every step of the way, every day. He had responded to Jenny's quiet "Thank you" when anyone came to the door. He had even been for a booster shot at the vet, and there had been no need for a muzzle. Jenny's family was, predictably, astonished. Equally predictably, they were now once more showing interest in Tim.

This is proof of what can be achieved through a combination of perseverance, self-belief, and faith. I felt proud that I had been able to help someone through such a tough process and, most of all, I felt proud of Jenny. I hope she provides a lesson to us all.

Acknowledgments

It has taken me the best part of twenty-five years to develop and then translate my ideas into the form in which they appear here. It has, I can tell you, been a long, slow, and sometimes painful process, one that I would not have completed without the help and support of a very special group of people. One of the true pleasures of finally completing this book is being able to say a heartfelt thank you to those individuals.

Firstly, I want to pay tribute to one of the most persecuted species on this planet: the wolf. This noble creature has taught me a great deal, not only regarding the behavior of the dog, but also about the failings of my own species. It seems paradoxical that mankind has almost wiped out this animal while taking its descendant, the dog, to its heart. And here I would like to acknowledge the dogs with whom I have shared my life and learned so much.

As for my fellow humans, I would like to thank initially the first people to show an interest in my ideas, those at BBC Radio Humberside. My thanks go to Maureen Snee, Blair Jacobs, Judi Murdon, and Paul Teage, each of whom has encouraged and helped me hugely. It was my work there that led to my appearances on Yorkshire TV's *Tonight* program. I would like to thank all the team on the show, but in particular my cameraman Charlie Flynn, who provided a professionalism that has blossomed into a friendship I value very much. In producing this book, I have been fortunate indeed in being published by HarperCollins, where Val Hudson's guidance and good taste have been priceless. The job of editing this book must have been an unenviable one. I would like to thank Monica Chakraverty for the wonderful way she has undertaken this task. I must also say thanks to Andrea Henry and Fiona McIntosh for their vital contributions.

It was my agent Mary Pachnos who brought me to HarperCollins. Her knowledge and expertise have been complemented by Tora Fost, Sally Riley, and the rest of the team at Gillon Aitken Associates in London. Without Mary, this book would never have come into existence. Without her initial interest, and the wisdom and wicked sense of humor that lightened the difficult days that followed, I would never have finished the task.

Aside from Mary, my greatest debt is to three men. The first is my partner, Glenn Miller, who has shown great patience and support in seeing me through the completion of this book. No one has played a more important role than Monty Roberts, the inspirational figure who changed my life. If I had not seen him at work a decade or so ago, none of the breakthroughs I have made would have happened. In the time that has passed since then, Monty, his estimable agent Jane Turnbull, and his dear wife, Pat, have extended me courtesies and kindnesses that I could never have expected. Thank you all, very much.

Finally I must pay tribute to my son Tony. Through often trying times, he has remained more than a son; he has been my most trusted friend and most reliable ally. Tony was the first person who made me believe I could achieve something worthwhile. "You can do it, Mum," became a mantra I repeated more often than I care to admit when times were tough. More recently he has become my colleague in carrying my work to a wider audience. He provided an invaluable guiding hand as this book was being written. I can't imagine a life without Tony. I dedicate this book to him.

The process of producing the Training Guide was in many ways more difficult than the *Dog Listener*. Once more, the fact that I have reached the finishing line is down to the professionalism and support extended to me by a special group of people.

First and foremost, I have to extend thanks to Val Hudson, Monica Chakraverty, Rachel Smyth, and Jo Wilson at HarperCollins. Their guiding hands have been invaluable at every stage.

I am also grateful to John Leach at Sunset & Vine for providing stills from my television series, and photographer Stephen Daniels for his hard work in capturing elements of my method on film. Thanks, too, to Sadie, my German shepherd, for the patience she showed during that long and demanding photo session.

Nearer home, the people with whom I work on a day-to-day basis have remained rocks, my closest allies. To the names of my partner Glenn and son Tony, I can now add that of Ali Powrie, our very special new dog listener. Most of all, however, I would like to thank my agent Mary Pachnos. What began as a business acquaintance has evolved into a valued friendship. She is my trusted shield—I would be lost without her.

Index